A BRIEF
TALE OF
OBESITY

A BRIEF TALE OF OBESITY

Dr. F. J. Fojo

Library of Congress Control Number:		2014902695
ISBN:	Softcover	978-1-4633-7881-3
	Ebook	978-1-4633-7880-6

This book was printed in the United States of America.

Rev. date: 10/02/2014

To order additional copies of this book, please contact:
Palibrio LLC
1663 Liberty Drive, Suite 200
Bloomington, IN 47403
Toll Free from the U.S.A 877.407.5847
Toll Free from Mexico 01.800.288.2243
Toll Free from Spain 900.866.949
From other International locations +1.812.671.9757
Fax: 01.812.355.1576
orders@palibrio.com
603558

CONTENTS

OBESITY CHANGING ITS SIGNIFICANCE

OBESITY. AN ENEMY WE MUST DEFEAT

To Isis: A challenge

To the millions and millions of overweight citizens, who except something more than words, advice, reprimands, impossible diets, peculiar exercises, medieval equipment, questionable medications, surgical mutilating interventions, and the relatively good intentions of advocates, experts in everything, trainers, merchants, doctors, scientists, researchers, and health managers...

"The women immortalized in Stone Age sculptures were fat; there is no other word for it. Obesity was already a fact of life for Paleolithic men or at least Paleolithic women."

R. Hautin (1939)

INTRODUCTION

All of us, at least those of us residing in countries with a high level of economic development, or relatively high, have the feeling that there's an increasingly rise in obesity in our society.

This perception is linked by a real increase, statistically demonstrated, of conditions that have, or seem to have, a close relationship with this increase in body weight: arterial hypertension, diabetes mellitus type II, hypercholesterolemia, dysmetabolic syndrome, numerous sleep disorders, psychological disorders, degenerative articular lesions, and even some types of cancers.

If the average body weight of human beings is increasing, and if there is a real correlation between this increase, which we popularly refer to as "fat", and the aforementioned disorders, then we must accept the fact that we are confronting a true epidemic, or even worse, a pandemic, with an exceedingly high economic rate, besides the loss of quality of life and early death.

In contrast, helping humans gain weight, which is no longer a common occurrence, or creating things that supposedly help them lose weight, has become a gigantic business. A considerable part of the food industry, as well as cosmetic, fashion, medical service, press, and even the arts, has a lot to do with

obesity, even though this unpleasant word is mentioned very little, unless used as an insult.

Let's say it in another way: we are constantly bombarded by an avalanche of items and products for beautiful people and to make us more beautiful, for healthy people and to make us healthier, for people in shape and to get in even better shape, but at the same time, we are getting fatter on a daily basis, and we are more unhappy with our physical appearance, with our phenotype, as a biologist would say, and even though in average we are living longer, we develop perspectives, conditions, and illnesses that make our stay in this planet less pleasant, and for many, very stressful.

It is openly known that this obesity epidemic is not only extending further, but it is also being noticed at an earlier age. Obese individuals are more and more precocious (like geniuses, but in larger numbers), and this fact has a threatening significance for two reasons.

The first reason is obvious; children and adolescents will be exposed at a much younger age to all the risks and complications that previously were only present at more advanced ages.

It's not the same to make your debut as a diabetic at thirty, than at sixty. It's not the same for our knees to carry extra weight at the age forty, than at ten. And we're not even mentioning all the financial costs that this type of precocity generates in a time of exorbitant medical expenses, tax deficits, and all types of cuts.

Yet the second reason is not commonly discussed; however, it has a greater potential risk for all of humanity: changes in epigenetic, which is different than genetics, may produce modifications in the genome (or in the way the genome operates), which are transmissible in short periods of time, and these changes, when produced before procreating, in other words, in the full reproductive stage, can jeopardize the offspring.

Although it's true that this is not absolutely confirmed, it is also true that obesity exponentially increases in everyone at the same time that obese children and adolescents also increase exponentially.

At the same time, after studying obesity from different angles for more than forty years, we coincide in that there is no single and coherent explanation for this mass phenomenon.

Unlike other illnesses attributed to a virus, a bacteria, or a phenomenon caused by immunological deficiency, obesity seems to be closely related to the very own genotype, social, and technological history of humanity. Common obesity is not just a polygenic condition, but it can also be classified as a sociological, economic, and historic aberration.

With this in mind, we decided to explore this history; the history of obesity and its multiple interactions with the evolution of society and mankind. This is not a diet book, or a book with weight loss treatments, even though, they will undoubtedly be mentioned; it is also not a medical description of obesity or a biological treatment for it.

Much less a detailed study of new bariatric or reconstructive techniques, it is simply a brief tale of how mankind has become bigger, thicker, and more obese. How and perhaps why, this obesity scenario is one of the most common in every city (and in the fields) around the planet.

Concisely, a brief tale of obesity.

WHEN BEING OBESE WAS "IN"

CHAPTER 1

¿Was Venus nurturing?

Only humans could have saved the dinosaurs.

A spacecraft, with a nuclear cargo, could have veered off the enormous meteorite that supposedly put an end to the existence of these enormous animals when it crashed against our planet, but... there were no humans to do it.

It seems ironic that with the bad press mankind currently has in regards to the destruction of species, it can occur to us to plan the hypothetic salvation of the gigantic dinosaurs with the employment of mass destruction weapons. But that's how we are. We'll let me then narrate, in a few paragraphs, the story of how we got here.

Between six and eight million years ago, when life had a very long history and tenths of thousands of different vegetal and animal species appeared and disappeared, a small group of mammals began to move throughout regions of eastern Africa, a continent without a name yet. Once in a while, these mammals could launch a stone and perhaps stand up on their two hind legs.

Remains found of these animals are scarce and fragmented; and investigators, paleontologists, and anthropologists, have named them Ardipithecus. In reality, at least four different types of these mammals have been discovered, but the most common is the Ardipithecus.

These animals, because they were animals, began evolving, which is very confusing for our existing understanding, and they made way, on one part, to the Australopithecus (afarensis, africanus, anamensis, etc.); and on the other part, to the Paranthropus.

For some specialists, the Paranthropus make up a different lineage and have no relation to the Australophitecus, but let's not complicate our lives right now. Almost every year, new discoveries are made and modern genetics may put order to this chaotic history. Maybe we can even clone one of these beings!

Let's continue then. A little less than two million years ago, the first homos appear: Homo ergaster, Homo erectus, Homo habilis, Homo antecessor, Homo rudolfensis, and some others. They all walked straight already, they had more or less large brains, and they manufactured a variety of stone instruments; however, they're not quite human yet, in the strict sense of the word.

Some of them left Africa, their crib, and their fossilized remains appear today throughout Asia and Europe. Some disappear forever and others evolve. The ones who evolved, gave way, approximately 200,000 years later, to two species which had to live together (we don't know if scrambled) for thousands of years. Even the first researchers got them confused sometimes. We are talking about the Homo neanderthalensis and the Homo sapiens. The first group, which knew and used fire, manufactured primitive tools and weapons, created art, and buried the dead, completely disappeared from the face of the earth about 25,000 to 30,000 years ago. The other species... that's us.

Whatever happened to the Neanderthals is a mystery. We do not have a satisfactory scientific explanation and we could actually be faced by a grisly thriller, since one of the theories in vogue, is by Professor Tzedakis, from the University of Leeds, who raises the possibility that the Homo sapiens, the Cro-magnum, meaning, us, ethnically cleaned out the Neanderthals completely.

If this theory is confirmed, monsters like Hitler, Stalin, and Pol Pot would simply become no more than incompetent apprentices.

Of course, there are other explanations: melting glaciers, the disappearance of their customary food, genetic degeneration due to mating between a few of them, epidemic diseases, and others.

Professor Svante Paabo, from the Max Planck Institute in Germany, has been able to obtain long DNA sequence from a Neanderthal found in Croatia (other researchers are in progress) that may prove some reproductive relationship between the Neanderthals and the Cro-magnums, which, if true, would explain some of the advances of the latter group, probably being more intelligent, but less adaptable to cold climates.

Whatever the case, the Neanderthals were extinguished and we remained to "dominate" the planet.

And the fat? Those first humans, let's say from about 25,000 years ago, had to survive in a very adverse and brutal environment, full of constant threats and crushed by constant climate changes.

Let me give you an illustrative example. A member of the special military forces of any developed country must endure a survival training program in the jungle. He is young and in excellent health; he is well nurtured and all his vaccines are up-to-date; he is equipped with a very powerful firearm, ammunition, a very sharp steel knife, matches, canned food, a water canteen and water purification tablets, night vision equipment, a compass, GPS, thermal uniform, bug spray, and a radio. If anything went wrong: a serious fall with bone fracture, being bit by a poisonous snake, the loss of equipment and guidance, he would only have to activate a MicroBeacon device and a support team would soon arrive in a helicopter to rescue him.

So, our homo sapiens from 25,000 years ago, was living in similar conditions, but with a few variants.

He had absolutely none of the aforementioned sophisticated tools, he was never nurtured, his protection from illnesses was almost null, his hands, and perhaps a pole or a stone, were most likely his defense weapons, and above all, nobody would come and rescue him, and in his whole life, from birth to death, which was quite often premature, he would remain in the same place and live under the same conditions.

Such living conditions would, little by little, prepare this man to fight for his own survival, creating neuronal connections that, in the long run, even in a geometrically increasing sequence, would turn him into a human being, endowed with intelligence, which we are today.

Abstract thinking, capacity to plan, combination of ideas about different aspects of life, symbolisms, and continual and unlimited generation of questions and answers turned that reactive brain into the human brain, computing (and spiritual) machines who no longer had any considerable rivals. And what did that translate into?

Well, in that jump, at first very, very slow, almost undetectable, and later, in increasingly fast pace, which hasn't ended yet, they created tools, weapons, means to cover and protect themselves, fire control, increased nutrition, and above all, transmission of information to their descendents.

Their genetic was also subjected to changes. Let's take a closer look at these facts.

Let's start with genetics. The nutrition of those creatures, who were not yet familiarized with agriculture, farming, and means of storage, was completely random.

It depended on scavenging or the discovery of a source of nourishment—a large and edible animal—that could be hunted. When this happened, a feast would occur! A quick and large banquet of meats and fats that had to be quickly eaten before it went bad or before other beasts or rivals discovered where the smell was coming from.

Those who had the capacity to store excessive fat in their own bodies, today we know in cells called adipocytes, could rely on some nutritional protection during times of abstinence, which would customarily come later. Those who did not have this capacity (having less fatty cells) were frankly at a disadvantage, and at the end, they would have to pay dearly for that.

Of course, this description is very graphic. Those men also ate produce, insects, some birds, roots, and leafs, which provided them with fiber and vitamins; some low-quality proteins and carbohydrates, which was quickly consumed due to the constant physical efforts of those days. However, and quite often, these foods could not be found in sufficient quantities and dentures and digestive enzymatic systems were not appropriate, in general, to process most of the vegetables.

In time, cooking changed this situation, and soup emerged, which was a gigantic gastronomic advance throughout the long road to human development.

But the ability to store fat continued to be quite important, and genetic lattice, not a single gene, but a bacteria of acting genes, facilitating genes, and "inactive" spaces of genome, which favored the accumulation of reserves, were becoming more common in men and women, of course, primitive humans.

This hypothetical mechanism that we've described here was proposed in 1962, by professor of Genetics James Neel, from the School of Medicine of the University of Michigan.

He called it "thrifty genotype". Ironically, Dr. Neel, who dedicated most of his life studying genetic changes produced by the atomic bombs in Hiroshima and Nagasaki, stated just before he died (in the year 2000) that he no longer was quite sure of his theory anymore, and that he believed that today's high consumption of empty calories is what brought about the high increase in obesity these days.

Maybe he's right, or perhaps his death, which occurred before the success of the decoding of the complete human genome (2001), prevented him from clearly seeing the complicated systems of the genetic-hormonal interaction.

In this day and age, many researchers believe that obesity is due to the association of a genomic strategy, a new way of identifying the old thrifty gene, as well as current bad eating habits and lifestyles. But we're getting ahead of ourselves.

And what do we know about the art and culture of those humans? Well, we do know something, not everything we would like to, but we have found sufficient evidence of their paint and sculpture skills in caves and burial sites. We are now confronted with humans who already have a symbolic structure, a purely practical way of viewing life and of explaining it. We are dealing now with the beginning of what would later become philosophy, religious ideas, and art.

By definition, prehistoric art must include all forms of creations from primitive communities: Africans, Mesoamericans, Asians, etc., but this goes beyond what we want to emphasize in this book. Therefore, we will only briefly refer to the cave paintings and the famous European Venus.

Upper Paleolithic pictorial art (between 30,000 and 10,000 BCE) is basically found in caves of the Spanish Cantabria region: Altamira, El Castillo, and Bustillo, and in the South of France: Lascaux, Font de Gaume.

It is an art considered appeasing for hunting, and thus, to attract food. Hunting scenes, where animals are fundamental (future food) and the hunter is secondary.

But what's truly impressive about their work, is the small statuettes baptized in modern times as Venus.

These are figures carved in calcite, serpentine stone, coal, ruminant antlers, mammoth fangs, hematite, clay, and other relatively common materials. They are small, almost always between 7 and 25 centimeters, even though some have been found, like the Fish Goddess (at the Belgrade University) that measures 51 centimeters. Invariably, they are all women. Three hundred of them have been found so far, although it may be possible that some private collectors have in their possession a few out of circulation.

They are not particularly from a specific region, and they're still being discovered, which demonstrates that their use, whatever it was, was of great value to that entire population. The most famous are the ones by Willendorf (11 centimeters, at the Museum of Vienna), Dolni Vestonice (11 centimeters, at the Museum de Brno), Grimaldi or Polichinela (8.1 centimeters, at the Saint Germain Museum), Lespugue (14.7 centimeter at the Museum of Man in Paris) and the Bajorrelieve by Laussel (48 centimeters at the Bordeaux Museum).

All, or almost all of them, make an explicit tribute to genocide obesity. Women with relatively small heads, without a face, many times without arms, with enormous hanging breasts, large buttocks (steatopigia), stunning hips, large bellies (they almost look pregnant), and very denoted and disproportionately big genitals.

Cellulite, even though not specifically represented, is there. A very interesting detail: they are never found in burial sites but in ruins of cabins and caves where people resided.

What are we describing? What were theses statuettes used for? We really don't know. Explanations are many and diverse.

Let's see: For many years, they were considered fertility goddesses, for human maternity as well as to attract abundant food, but we can't find any single scientific or historic proof that supports this. In 1996, anthropologist McDermott speculated with the possibility that they were sculpted by women as a way of self-portraying themselves; but he also couldn't provide evidence for this.

It has also been said that they were a tribute from men to their women's fertility, which is also pure speculation. What if they were distant forerunners of Playboy and XXX films? This is a joke, but it has as much value as any other hypothesis, and if so, they would constitute proof that those Paleolithic men admired and desired voluminous substances and morbid masses (up to what point was masturbation common? We don't know this either).

The reality is that thin women had very bad press at that time, which makes sense in light of the thrifty gene.

What better mother and woman than the one who brought with her, in her exaggerated adiposities, the necessary reserve to survive times of deprivation and preserve, in her large breasts, her young ones with life.

And later? Later on came agriculture, farming, and gastronomy. Neolithic Revolution, as referred to by Australian archeologist Vere Gordon Childe (1892-1957).

A scented food and an exhilarating experience

If you walk by an area near the Iceland Sea, and you smell the strong odor of ammonia—describing it as fermented urine is a bit rude—don't think there's dumpster nearby, no, it is almost certain that they are processing Hakarl shark meat.

This shark cannot be eaten fresh since it has a large amount of urea in its flesh (it has no kidneys), making it highly toxic.

In the curing process, which takes about three to five months, almost all of the urea is removed from the meat. Holes are dug in the rocky ground, as far away from residential areas as possible, and that's where the "smelly" process is carried out.

Icelanders offer foreigners the Hakarl, with a liquor called Brennivin, but they can rarely prevent their guests from throwing up; always urging them, in-between laughter, to cover their nose and swallow fast, but not even like that!

San Nak Ji is something else; it's a whole new limited experience.

It's just the common and ordinary octopus, but... alive and kicking. The idea is that they place them in front of you and you cut their tentacles, which are desperately kicking, and then you must swallow them without chewing much.

They're delicious, even though they don't give you much time to savor them.

Statistics show that every year, at least six deaths occur in Japan among fans of this dish. The deaths occur from choking with the moving tentacle in the trachea.

Most of the deaths are from foreigners, who with lots of enthusiasm and lack of knowledge, engage in this limited gastronomic experience.

Would you like to try it?

CHAPTER 2

The decadence of the predators

For us, modern people, a millennium is a long time, and we have a really hard time understanding how Homo sapiens, living in those conditions of dreadful shortages and limitations, which we call "prehistoric times", could have lived for two hundred thousand, three hundred thousand, or even more years, many more than those 7,000 years of which we only have some fairly reliable historic information that we are generally proud of.

Homo sapiens, which is how we scientifically call ourselves, had to first make something of himself; then, he had to impose himself to nature, survive and prevail, endure several glacial periods, defeat and perhaps eliminate other hominids (such as the Neanderthals), disperse himself throughout the planet, coming in successive waves from the deepest regions of the African continent, invading, surreptitiously, Eurasia, crossing over the Bering Strait, which back then was not a strait, but an irrelevant land bridge (of course, it wasn't called Bering!), extending himself through the Americas and even learn to navigate in order to populate the islands of the Pacific.

We grew up hearing about the famous "missing link"; yet, our human forefathers lived with it, or with them, for thousands and thousands of years. It's a pity there was no writing or cameras back then.

This very long period can only be researched in fragments, relying only on a few fossils and archeological sites, which in time may have been neglected or ransacked.

After this odyssey, and once established as the only creature capable of creating wealth, at first very meek, and of generating what we would later call culture, mankind begins to settle down, to become sedentary.

What was it that moved him to change his routine of tens of thousands of years? Just like all other questions in this topic, we don't have a concrete answer here, even though there are plenty of hypothesis and theories.

In truth, hunting, gathering, the vestige of scavenging, and nomadic behavior visibly declined to give way to an accelerated and unedited process of territorialism. Agriculture, the domestication of animals, the division of work, the emergence of food overindulgence dramatically changed the aspect of this man and turned him, in just a few centuries, into.... us.

This process was not lineal and perfect. It was full of highs and lows, partial setbacks, catastrophes, and in some cases, even fatal tragedies. It wasn't a synchronized process either; when some groups were already establishing settlements that could be called "cities", others were still quite backwards, or even at the verge of extinction, as we can see nowadays in some tribes in the Amazon. The extinction process of the Mayan society in Central America, a process which is far from being explained historically, is very illustrative of the highs and lows and setbacks of the social evolution of a civilization.

And what happened to the people who were a little overweight? Well, they continued existing, but there were some important changes in perception.

With permanent settlements and the development of stable societies, small clans and wandering tribes, formed by dozens or hundreds of human beings, grew and diversified.

Slavery made its appearance, product of the intelligent observation that a slaved enemy rendered services that a dead enemy could not, provoking in turn predatory wars and plunder.

In order to take slaves and control them, warriors had to come into view, who in turn were developing a pyramidal military structure, topped with the strongest ones, later called chiefs, generals, kings, emperors, etc.

And with these chiefs, who now only work once in a while, came leisure, luxury, and inequality.

Food products were multiplied with the establishment of agriculture, farming, and initial storage. This brought about a new form of exchange, which in time would be known as commerce. Eventually, this also brought in currency, one of the most transcendental inventions of humanity.

Those who had to work all the time did not eat well. The majority couldn't even dream of gaining weight. However, the wealthier groups, with access to food excesses and free time, began to enjoy the pleasures of diversification and sophistication of food. Banquets, buffets, and orgies were invented.

Many women kept themselves overweight in order to feed their children, or even other women's children. Their fat reserve came in handy in times of famine and devastation, which occasionally returned and brought in wars, or when unforeseeable weather conditions occurred that could not be prevented.

And so a new type of fat emerged, above all for women. Fat as ostentation of wealth. Not only did they show off their ornaments, jewelry, and flashy weapons, but also their flesh, as proof of power and image, evident of the possession of slaves and leisure time.

Lifespan also increased relatively, since there was more stable and better quality nutrition. They were eating many of the foods that we still eat today: cereals, bread, honey, fruits and vegetables, sundried fruits, the first alcoholic beverages, milk, fish, poultry, a variety of meats, and more.

With extensive and easier ways of making and maintaining fire, and the employment of some kitchen utensils, mostly made of clay, emerged the art of cooking; gastronomy, which gave food a new dimension, now not only as a necessity of life, but also as taste and pleasure.

A plump man, wrapped in flesh, must be the chief, a nobleman, or someone close to some form of power. A woman wrapped in flesh was enticing and beautiful, because she surely worked little or nothing at all, and could dedicate more time to herself and her man.

There was still a long, long road to the ritual of starvation diets and gym tortures.

A plump man stuck in a chimney

Do you know who Donner, Blitzen, Comet, Cupid, Prancer, Vixen, Dasher, Dancer and Rudolph are?

If your answer is that they are the names of the reindeers that pulled Santa Claus' sled, then you're ready to become rich playing a television game show.

This ruddy and fat man, owner of the reindeers and sled, is well-known almost (almost?) everywhere in the world, not with the same name, though.

If you are Swedish, you will know him as Jultomtem; if you were born in the Netherlands, he will then be Sinterklaas; or Julemanden if you are from Denmark. In Russia, he is Ded Moroz, and in Romania Mos Craciun; In Italy, he is known as Babbo Natale, in France Pere Noel. In Portugal, he is known as Pai Natal, but in Brazil, where they also speak Portuguese, they call him Papai Noel.

The Germans call him Nikolaus, and the Spaniards, Argentines, and Colombians Papá Noel. Chileans, in a loving way, call him Viejo Pascuero, and Costa Ricans Colacho. But if you are Cuban, you surely call him Santi Clos, even though if you reside in Florida, you'll probably simply call him Santa.

In the end, why go on? One way or another, we are talking about one of the most famous chubby man in the entire planet.

But who is he?

If a time machine could take us back to New York City, on December 25th, 1860, you would surely feel some frustration: we would not be able to see the familiar image of Santa Clause anywhere!

It would probably be snowing, carriages clattering their rattles, shopkeepers calling out their merchandise, and the organ playing some kind of Christmas song, but chubby Santa, dressed in scarlet red, with his famous "ho-ho-ho", would not be found anywhere.

But this need would soon be met.

In Harper's Weekly edition of January 3rd, 1863, the most important and circulated magazine in New York City in those days, plastered on the front cover, is the drawing of Mr. Thomas Nast. An older, potbellied, bearded man, wearing a heavy suit and a pointy cap on his head, approaching a group of northern soldiers (we are in the midst of the Civil War in the United States), offering them gifts and candy.

The fat man is sitting in a sled, and in the back you can distinguish two reindeers. In the bottom of the drawing, there's a metallic arch with the inscription "Welcome, Santa Claus".

This is the first published image of the obese character we know as Santa Claus.

And who created him?

Well, the creator is also quite a character!

Thomas Nast was a poor German immigrant who arrived in New York City at the age of 6. At 15, he earned a living as a recorder and cartoonist, and at 18, he worked as a journalist, becoming more influential and important for Harper's Weekly.

He was a personal friend of President Lincoln and President Grant. He ruthlessly attacked the southerner secessionists; he was with the troops of Garibaldi; confronted with a pen, and almost lost his life, against William Tweed, the Tammany Hall Boss, successfully putting him in jail; he was a friend of Mark Twain, and on occasions, saw himself entangled in shady businesses; in short, he was quite a character.

But his combats and adventures are history.

Santa Claus is not, he prevails.

CHAPTER 3

The first civilizations

Mankind begins to scatter civilizations. From this point forward, we can rely not only on archeological and paleontological findings, but also historic references made by the own inhabitants of those communities, who perhaps had something to celebrate or tell.

Now, we get to learn how they identified themselves, as well as their gods, governors, enemies, communities, and sceneries.

Carved stone steles, clay tablets, written papyrus, paintings on vases and glasses, colored ceramics, architectural constructions and many other forms of expression, tell us that prehistoric times, to a greater or lesser degree, has come to an end, and history has begun.

A few remains of the Hassuna civilization were found to the north of the Tigris and Euphrates rivers, in the territories that today occupy Iraq, Syria, Iran, and Turkey. There's not much left on the land that it once occupied: thousands of wars and devastations have done their job, but you can still find pieces, very little indeed, in some museums and universities, that tells us bits and pieces of history. Painted receptacles and murals, kept at the Bagdad Museum, describe their agriculture and farming.

In the mural, we can see an ornament, with very fine details, that shows the daily work carried out in a stable: the milking, the caring of the animals, the transferring of milk from one receptacle to another, etc. And curiously, very nourished human figures, about eight of them, are distinguished. Even though they were probably slaves, nutrition doesn't seem to have been a problem for them.

The Halaf, Samarra, and Ubaid cultures emerged later, and sometimes were integrated. They built houses, small temples; they buried the dead, and used irrigation for their crops. And just like today, eating and battling were the most common activities.

These human clusters were named pre-Sumerian, since subsequently, around 4000 BCE, the Sumerians made their historic appearance, building a civilization exactly as we understand it today. We don't see overweight people painted on their steles, only militaries, conquerors and the defeated, slaves, as well as those who died in battle. However, small sculptures of obese women, generally seating and endowed with enormous breasts and thighs, were found.

An especially interesting piece is a terracotta figure found in Catal Huyuk, measuring 6.5" high, which emphasizes the genitals in quite an obscene way, at least for our current standards. Was this a tribute to procreation or simple lust; a protecting mother or a porn star?

The Acadians came almost a millennium later, and they settled more to the south. However, one of its kings, Sargon the Great (2296-2240 BCE), completed the unification of Acadia with Sumerian.

This is how the Mesopotamian civilization is born: the Babylonians, named after their beautiful capital. Babylon, the one with the famous hanging gardens, was a great city, full of life, commerce, pleasures and food. It's a pity to see, in documentaries about the Iraq war, its deterioration and scarce remains.

To this day, a Babylonian statuette has survived, which is probably the figure of the first well-represented fat person in history. He wasn't a king, whose image artists usually improved, but a simple laborer with a sedentary job, perhaps responsible for keeping the sovereign accounts, benefiting from residing in the palace and eating hot meals every day. His plump and serene face allows us to enjoy his good life.

Around that time, the Harappa civilization was already growing around the Indus River, along with the first Chinese Empire, the Shang Dynasty, at the margin of the Yellow River, and very close of the Nile, the Abydos palace complex, the first important Egyptian settlement.

Today, we know that around those dates, other corn-based proto-civilizations were being born in Mesoamerica, and a few centuries later, in the Gulf Coast of Mexico, the Olmec civilization, still involved in mystery and legend.

Dispersing, conquering, and complicating things is a substantial part of human condition; dozen of thousands of years were needed for a few group of primitive men to settle in fertile and irrigated locations to construct small communities. But once agriculture and animal breeding, with the immeasurable help of fire and growing metallurgy, gave way to the first great wave of development, civilization was made present almost everywhere on Earth, with the emergence of towns, cities, kingdoms, dynasties, cultures, wars, and politics, which unfailingly require appointing temporary periods, classifying eras, and giving chronological order to this sum of events and facts that make up what we call history.

To simplify and stick to the subject of our book, let's quickly review a few centuries ahead. The Hittites emerged, a belligerent and imperialistic community, stopped by the Egyptians in the almost mystical battle of Kadesh. Its primary goddess, the Great Mother, was a woman with abundant meat in her bones and very prominent sexual organs, which they adored, like the Moon.

The Phrygians, creators of the Gordian knot, cut off much later by Alexander the Great, invented the legend of King Midas, the man who came up with the perfect diet, because after insistently begging the goddesses to allow him to turn everything he touched into gold, he ended up dying of starvation.

The Aqueous, predecessors of the Greeks. The Cretans, also entwined with the first Greeks, left us the images of their dressed women, but with their voluminous naked breasts, a fashion that would later be stylish again, 3,000 years later, with the "topless" trend. The Semites, with their two most important branches, the Phoenicians, merchants and transporters of merchandise through antonomasia—some comedians would say they invented UPS—and the Hebrews, very much the merchants, like the Phoenicians, but much more cautious and given to accumulation of money.

The Chaldeans and the Medes, discussed by some scholars as a state, very close to being crushed by the Persians.

The Iranians, protectors of the path of Khorasan, a long road through which they transported, among other things, merchandise, food, and delicate gastronomic specialties. The Anatolians, and many more.

All these civilizations, and many others that have disappeared or have transformed themselves through integration or absorption, shared a common socio-economic structure: slavery, which together with the power of agriculture and the domestication of animals, boosted them up, at the moment, to historic stardom.

It has been calculated that a grain of planted wheat produced between 60 (in low seasons) and 80 grains of new wheat, which allowed for the emergence of bread, the food of the multitude, fuel for the optimum performance of slaves and warriors and social paradigm.

This nutritional abundance opened a new path to art, commerce, politics and the rudimentary forms of science.

Being fat or overweight was attributed to those who had no need for strenuous physical activity: governors, scribes, palace bureaucrats, noble women, or perhaps those women who performed limited muscular work in palace harems.

It is still uncertain whether the idealization of female sexual attributes—breasts, buttocks, thighs, and vulvas—

exaggeratedly thick for the patterns of those days, and constantly represented by artists, had a religious basis, or, on the contrary, perhaps they intended to satisfy the sexual lust of men who could only find extremely thin women due continuous labor and hardships in food shortage.

What is very certain is that what doesn't abound seems more attractive.

Foie Gras

In an Egyptian tomb excavated in Saqqara, dated approximately to 2500 BCE, a low relief sculpture was found representing geese fatteners, using a technique very similar to the one used today.

The procedure is as follows: the birds are clutched by the throat, and the food is forcefully stuffed into them by opening their peak, sometimes using a hollow reed as a gadget.

Of course, nowadays, we use rubber or Teflon tubes to make the process quicker, as well as more "humane" for the geese.

From Egypt, the art of, let's call it "fattening" of the geese to later eat its fatty liver, was spread to Greece and later Rome, like many other things.

It is said that Marco Gabio Apicio, a Roman senator and epicure, lover of good cuisine and other flesh pleasures, introduced the use of dehydrated figs to force feed the geese, with which the quality of the liver substantially improved.

Let it be clear that the quality of liver improved from a gastronomic point of view, because for the bird, this statement may be a bit uncertain...

However, we must recognize that these birds, with the power of flying long distances, sought their own misfortune, because they store their energy reserve (calories) in their liver, which allows them to migrate, from place to place, without eating, which was detected by the Egyptians, maybe by coincidence.

In the Middle Ages, it was the Jews who preserved the production of Foie Gras (literary fatty liver in French), due to the kosher prohibition of using lard to cook.

Later, the French, those magnificent gourmets, made geese liver their own, artificially increasing their size, or Foie Gras. Since then, no one has been able to take it away from them, even though many have tried and globalization has significantly changed the rules of the game, from the gastronomic to the commercial point of view.

Industrial production of Foie Gras includes intubation of the animal's stomach in order to be overfed, the impediment of flying and even moving, so they won't use up any of the fat stored in their hepatic cells in energy production (besides doing the whole process as fast as possible) and sedation (or anesthesia) prior to sacrificing them in order to avoid stress for the bird, which has an immediate impact on the condition, presence, and texture of the liver.

Some producers have defended the entire process by alleging that it is even less traumatic than what the cattle goes through at the slaughterhouse. Makes you wonder...

Foie Gras is illegal in many countries (about fifteen of them around the world), but it continues being produced and consumed; there is even an international black market for this product.

The legitimate and purest kind is quite expensive, but you can find cheaper ones.

CHAPTER 4

Sand, Queens and Pyramids

When we think of Egypt, back in the time of the pharaohs, we can't picture thick or heavy people. Large extensions of sand or lands swamped by the floodwaters of the Nile, men bent over with the blazing sun hitting their backs, cultivating the land or pulling huge stones with thick ropes to construct pyramids and tombs, warriors driving carts enveloped in clouds of dust. Strenuous work and an intense and short life.

In reality, this is how most Egyptians lived their lives on a daily basis... with few exceptions.

Hemiunu was a fortunate man; he was the cousin (Egyptian lineage is difficult to understand, even experts) of Pharaoh Keops, the one who ordered the construction of the great pyramid that carries his name, located in Giza, very close to the current city of Egypt.

Anyhow, Hemiunu was the architect who directed the construction of the pyramid by orders of his cousin, the Pharaoh, and that meant a great honor, since the Pharaoh put in his hands and in his talent as constructor, the responsibility of making sure his body, his mummy, and his personal wealth weren't violated or ransacked throughout his long and venturesome journey to the afterlife.

We don't exactly know what perks Keops gave Hemiunu. We assume many, since Hemiunu's wealth was immense and obvious, but we did know that Hemiunu decided to award himself in life by constructing his own statute, which we can admire in the European museum of Hildesheim.

We recognize in it a relatively young man, with a round face and very well shaved, fat, seating over his thick buttocks, covered by a flap and, most striking, with breasts that appear like a woman's due to its emphasized adiposity.

Doctors would say that Hemiunu suffered from gynecomastia, which means a disproportionate growth of the mammary glands, like that of an adult woman, but quite uncommon in men. Hormonal disorder or simple obesity? We don't know, but what we do know is that Hemiunu felt very comfortable with his weight, and with his breasts, since he could easily have covered them with a robe, typical raiment of that time, but he didn't.

The reason for Hemiunu's obvious pride is simple.

Fat, adiposity, prominent bellies, and thick buttocks were for the Egyptians, and for almost every civilization of that time, an unequivocal indication of wellbeing, wealth, power, and good connections.

And what a connection Hemiunu had with the royal family!!

The Pharaohs, on the other hand, didn't display their stoutness, but we assume they didn't have to; their divine attributes, very clearly exhibited in sculptural and pictorial representations, were enough to squash the rest of the mortals, which was not the case with the pharaohs' subjects and courtiers, who, despite being powerful, had to seek for more earthly ways to show their status.

But there were fat pharaohs, and some very fat, such as Akenaton, a sick man, and perhaps a little crazy, or maybe a revolutionary mystical character among others; or Ramses III and Amenhotep, who are pictured very fat in different low relief sculptures.

But the most astonishing Egyptian statute (or of that time, we are not very sure of its original origin) was that of the Queen of the Land of Punt.

We don't know if this small masterpiece of antique sculpture was sent to Egypt as sort of a portrait of the queen so people could know what she looked like before her arrival, or if it was created during her visit, which was quite prolonged.

We can enjoy a young woman—back then they were all young—with black African traces, very attractive, and with hair completely in braidlocks, a wicked smile and a body similar to that of batrachians: a big belly, a roll of fat under her breasts, and imposing thighs and buttocks.

The curious thing is that, despite the fat, the queen's body still had some charm, certain sensuality, something like a model in a plus-size magazine or a Caribbean performer.

Still a question remains unanswered. Given the disturbing beauty of Nefertiti, proudly slim and very sophisticated: Was the queen of Punt a woman proud of her fat and her disproportionate physical attributes, or did she act as a seductress with her female wisdom, or even more intriguing, did she know that all Egyptians, deep inside, liked fat women and she wanted an audience before making her appearance? We don't have this answer, but it's still an interesting reflection.

Underprivileged Egyptians, which were the majority, lived basically on bread and beer, a drink they were able to produce very well, and of which they had different types, flavors, and textures.

The wealthy had abundant food, and quite balanced, too: poultry, fish, produce, and milk, but they also consumed bread, so it may be possible that bread was the cause of their bad teeth, which the poor and the rich suffered alike.

It has been demonstrated that the bread placed in their burial sites, as an offering for the long journey of the deceased to the afterlife, contained large amounts of sandstone, almost certain as consequence of the way they grinded the cereal. Egyptians' dental disorders were constant and unavoidable, and unfortunately, these caused severe infections that directly, or through renal or other complications, lead to early death. Cavities, dental abscesses, and loss of teeth were a curse for the Egyptians. They could be very proud of their monumental tombs, but their dentists were a disaster!

Egyptian art—and it is very important to point out that they did not consider it as such, but a simple representation of their daily reality, something like the social pages in a contemporary newspaper, but inscribed in stone—can teach us a lot, even about the coming and going of our body image and our current attitude towards it.

In paintings found in the burial site of Djeser, around 1500 BCE, we can see a relatively thin young woman fixing an older woman's hair. This young woman, who bends forward, has three "floating devices": three rolls of fat in her abdomen that she carries with much elegance and unconcerned.

Of course, we can allege that liposuction wasn't invented yet, but the dignity and pride with which they showed their bellies, double-chins, and flabby stomachs make us think that plastic surgeons weren't' needed there, if there were any back then.

The pharaohs' empire deteriorated throughout the centuries and Egypt ended up as a Greek colony, then Roman (it's said that Cleopatra was also a woman with meat in her bones), Ottoman, and then English.

Today, it is a great independent country that drags the consequences of populist governments, military dictatorships, and flirtations with antidemocratic groups, but that unexpectedly and arduously pulled itself up, especially thanks to its youth, to try to recover that ancient historic and cultural splendor.

Nutrition and gastronomy

Nutrition studies the biology of physiological needs and the use of foods by the organism.

When we say that proteins or vitamins are indispensable for the preservation of life and good health, we are not analyzing the form (foods) in which we incorporate these proteins and vitamins into our bodies. In other words, we are analyzing aspects of nutrition but not gastronomy.

But when we speak of preparing a balanced menu that contains the necessary proteins and adequate vitamins, presented pleasantly, then we're talking about gastronomy.

Gastronomy relates to human culture and only human, since no other animal cooks, seasons, mixes, or prepares any form of food. In any case, humans are the ones who created gastronomy for animals, and humans are also the ones responsible for making their pets overweight or obese, but that's another story.

Gastronomy, by definition, is the form of manipulating, cooking, and presenting food in a more pleasant and possibly tasteful manner.

Here are a couple of examples: When a cow eats grass, it is nurturing itself, but that has nothing to do with gastronomy.

When we eat a good steak with fries, we are also nurturing ourselves, but at the same time, we are enjoying our mom's gastronomy or that of an expert chef.

Nutrition is a physiological need that is scientifically studied and understood; gastronomy is an art that makes our life more pleasant.

Gastronomy is part of nutrition, but it is not all nutrition.

CHAPTER 5

Beautiful and balanced. The Greeks.

Some two thousand years before our era, numerous Indo-European migratory waves arrived to the nearly uninhabited basin of the Aegean Sea. They brought with them their customs, their languages, and a huge advancement of military significance: the horse.

After blending in with the sparse inhabitants of the region, already settled there for two or three millenniums, they created, little by little, a new civilization; the Greek, which at least, on the cultural and philosophical point of view, would change the world, the way we look at it and understand it.

The Minoan culture emerged in the Island of Crete (by King Minos). We have already referred to this civilization through the paintings of its beautiful and opulent topless women.

With a strong and centralist government, the Minoans established colonies in the continent and extended commerce, sometimes by force.

They practiced what we would call today "sports", and invented a very particular form of bullfighting, whereby its athletes, of both genders, practiced gymnastics on the hump of wild bulls, a very important animal in this civilization's mythology. The eruption of the volcano in the Island of Thera may have had a lot to do with the decadence, and subsequent disappearance, of this mysterious civilization.

Then the Mycenaean civilization emerged. They lived in the continent and substantially benefited from the Cretan decadence. They were good seamen, and why not say it, better pirates. They blended in with the Achaean civilization and were the first inhabitants of those truly Greek lands, in the sociological sense of the word.

They fought the Trojan War, allegedly because of a beautiful woman, Helena, and they won with weapons and their shrewdness.

The Trojan horse, whether it existed or not, left an eternal paradigm to define treason, lead by intelligence.

With the decline of the Mycenaean civilization, Greece quickly disintegrated, which gave way to the emergence of small, miniature kingdoms; embryos of the city-states, dominated by local governors—kinglets—who were actually the owners of those small towns and a few pieces of land around them.

The Dorians, Ionians, and the Aeolians, names that may seem familiar because of their columnar architecture, came from these impoverished origins.

During this historic period, the Greeks completed the colonization of all the territories surrounding the Aegean

Sea, turning it into a particular type of lake, which did not prevent the Felicitous, a true Mediterranean power, to almost completely dominate maritime commerce, due to the extreme fragmentation of the city-states and their intestine clash (cases like the Trojans or Achaeans were quite common, but they did not reach the fame of this family dispute between gods and men).

This historic period was named, perhaps a bit unfairly, the "Dark Ages", probably because of the comparison with what would come later: Sparta and Athens.

In a way, Sparta was an artificial state. We don't even dare think of a fat Spartan, among other things, because he would be expelled from the city! Hollywood made the military capacity and the incredible degree of self-sacrifice of the Spartans trendy again (300), which is true, but at the expense of eliminating newborns who were not "suitable enough" (according to the historic parameters that are not quite clear), the discrimination of women, the weak, and why not, the "obese", the intimidation and eventual destruction of the family and slaving or death of foreigners who had the bad fortune of accidentally ending up there or being taken prisoners.

Leonidas, the proud King who died defending the narrow passage of the Thermopylae (Hot Springs), together with his 300 warriors, allowed his men to liberally eat breakfast on the morning of the final battle against the enormous Persian army only "because they would eat that night in the underworld".

How could there be fat people among them?

Athens was a different story. It was also founded as a militarist state and actively participated in the defeat of the Persians, together with Sparta—let's not forget Marathon, Salamis, and Platea—but it evolved towards a democratic government, in part thanks to the efforts of a great politician and legislator: Pericles, and some part-time warriors, such as Miltiades.

This does not necessarily mean that Athens was wonderful! Slaves worked hard so that citizens could have a place to discuss and make laws in a public plaza, and their leaders had an emphasized imperialistic approach with foreigners, which lead them to engage in numerous wars and conflicts, which throughout the years accelerated the Athenian decadence.

Its story was a long succession of military victories and defeats, coup d'états, tyranny, and democratic governments (limited democracy only to Athenian citizens), but its philosophical, cultural and artistic legacy was immense.

What would have happened if Xerxes, son of Darius, would have triumphed and destroyed the Greek civilization, as was his stated purpose? Better not to even think about that!

Plato, Socrates, Aristotle, Anaximander, Thales, Pythagoras, Heraclitus, and another endless list of philosophers, were the men, and a few women, who discussed about any topic in their plazas and in their homes. Sculptures, who made a mark on us with their pattern of beauty, in the way we see others; top architects, scientists, lawyers, legislators, physicians, and dialecticians.

Nowhere, and in no other era, except perhaps for some moments of Florence Renaissance or the cold era of the

Manhattan Project (to quickly construct the atomic bomb in order to end World War II), have we seen so many men of intellect and superior wisdom gathered together in just one place.

Men who knew how to live and who enjoyed a good conversation as much as a lavish dinner, or a visit to the brothels, female or male, from the Pottery neighborhood.

Greek sculpture, which we mainly know thanks to later Roman reproductions, has constituted the parameter to measure the perfection and beauty of a human body, even today.

It is interesting to observe how, in this day and age, perhaps as a rebuttal effect to the prevailing obesity in post-industrial societies, many have abandoned this ideal and harmonic body, in order to try, at the expense of endless diets and exercises, to maintain an extremely thin figure, sometimes even scrawny.

For the Greeks, the word "diet" meant a lifestyle, which included, besides healthy food, physical activity. On the walls of the Temple of Delphos, is written: "Exactness is beautiful", "You must respect limits", "Hate insolence (hybris)", and "Never too much". So pragmatic for life!

The statue of Athena, by sculptor Fidias, one of the most balanced and physically healthy women of the entire history of art, would not be able to find a job today as a runway model! Her strong arms, quite thick but very well sculpted, and her rounded neck, would be criticized today as that of a woman with an obvious weight problem.

The same would occur with Capitoline Aphrodite, Apollo Belvedere, with her lower abdomen rolls; the armed

Aphrodite, by sculptor Polycleitus, which we can see at the National Archeological Museum of Athens, is ready for a liposuction of the hips and lower abdomen; the child holding the goose, by Boethius of Chalcedon, frankly overweight; and, probably one of the less balanced sculptures, despite being the most realistic of all Greek sculpting art: Apollo, the Hunter, at the Britain Museum, already showing—please take into account he is a boy—a growing belly with an android characteristic.

And the Venus of Milo, perhaps the most perfect of all, but with her growing, and to my understanding, very delicious, little belly. Let's remember that Greek philosophy established, as fundamental principle of its art, to humanize the divine and deified humans, and the greatest artists followed this principle.

Around 350 BCE, a Greek writer, Archestratus, wrote a book called "The Life of Luxury". Perhaps the first culinary recipe book ever published. Even current Greeks, many without knowing the author, use these recipes in their daily menu.

We owe the Greeks many truths that we take for granted today.

The concept of interior beauty conveyed outside through serenity and friendliness. The proportion between the parts, as an expression of harmony in what is beautiful. The "kalon", a concept that would explain the frustration of many people who spend their lives in a gym or in an operating room and do not feel loved: is simple, kalon is "what you love" and what you love attracts, it doesn't matter if it's trendy or in fashion. The concept of purpose, expressed very well by Socrates: "Beauty for the race is ugly for the battle, and beauty for the battle is ugly for the race."

Limits, established by common sense, are also a Greek concept that sadly we constantly forget in so many aspects of our daily lives. We still have a lot to learn from this wise and very relaxed civilization.

Being omnivorous

An omnivorous animal eats everything.

People (humans) are omnivorous mammals because they are physiologically capable and adapted to eat everything.

When we say "eat everything" we are referring to everything that is edible. It's true that there are many limitations—there are plants that are not digested by the human digestive system, for example—but then gastronomy comes around, and through cooking and a variety of other techniques, it makes mankind even more omnivorous.

However strange it may seem, most animals are not omnivorous.

The cow, for example, eats only grass, and the reason for this is that its digestive system and its enzymatic systems are not designed to digest, let's say... meat.

That's why a cow is an herbivorous ruminant, but is not an omnivorous.

We, humans, are omnivorous mammals, because we can eat and digest meat, fish, shellfish, eggs, almost all vegetables and a myriad of other foods, even many that are not natural but industrially created.

But besides that, we have invented cooking and culinary preparation, in other words, "gastronomy", and this extended almost infinitely our possibilities of nourishment and enjoyment.

Sadly, it also extended the possibilities of gaining weight, which is the topic of our little book.

The only other mammal in our environment, that is completely omnivorous, is the pig.

CHAPTER 6

The mind coaches

Around 477 BCE, in a small town named Pava, very close to the foot of the imposing Himalayan, a blacksmith from the village ceremoniously invited a revered old man, a bit overweight, to a feast that included pork and other delicious dishes.

A few hours after the feast, the old man—who asked to be taken to a bed made of hay, under a beautiful blossomed tree—died, allegedly from indigestion.

This old man, whose name was Siddhartha, but who later was known throughout the world as Buddha, or simply Buda, which means "the enlightened", already had thousands of followers when he died.

He was revered by his faithful followers, and, something very curious, he was not buried where he died. He was actually dismembered and parts of him divided throughout many kings, local kinglets, and members of the nobility, who placed these parts in small hills constructed for that purpose and known as "Stupas".

Buddha did not create a religion, not even he was a religious man. He created, or described, a way of life, a series of ethical and moral criteria, which would allow a person of strong will and self-control to reach a status, known as "Nirvana", where he would never have to reincarnate again, and therefore suffer, with successive reincarnations—linked one with the other—the pain of this world, which in reality is already hell.

Nirvana is not obtained immediately; it is a long and hard process that requires a lot of patience and many lives. There are nine enemies that oppose the attainment of Nirvana: Discontent, hunger and thirst, sloth, doubt, weakness, pride and fame, hypocrisy, voluptuousness, and yearning. There are other pitfalls to overcome during the journey to perfection with the nobility of eight paths: Decision, action, correct lifestyle, belief, thought, hard work, words, and meditation.

Explaining the entire philosophical, ethical, and mythological content of Buddhism, in just a few words, seems quite an impossible task. A few days before the death of the master, his closest disciples were already fighting, among each other, for his legacy, and the division of the sect began, which actually hasn't ended yet.

Books and articles about Buddhism can fill an entire library. A Tibetan monk, completely bald and dressed in an orange tunic, enduring cold temperatures and repeating the same mantra until exhausted—a prayer written over a small mill that continuously rotates—is a Buddhist.

The King of Thailand and his court, which worship Buddha but enjoy western comforts, are also Buddhists. And an

American executive who drives his BMW to his yoga class, and changes his designer suit for his sweat pants, also designer, aspires, or believes, he can be a Buddhist.

But, what did the Buddha eat? Indian gastronomy has not changed much throughout the centuries due to the frequent integration with invaders and colonizers, or simply the interchange with a variety of neighboring communities. However, it always had an enormous variety of culinary styles, and much localism, in its flavors and presentations.

Seasonings, which subsequently had a lot to do with the discovery of America by Columbus, were essential components in Indian cuisine. As well as bread, mainly made with whole wheat flour—the "atta"—rice and a vast variety of vegetables, besides milk and their derivatives.

In coastal areas, a lot of seafood is consumed, but towards the interior of the enormous country, especially in the northern mountains, precisely in the territory where Buddha lived, nutrition is basically vegetarian, which makes us think that perhaps what killed the enlightened was excess pork meat.

In any case, Hindus have usually not expressed their religious imageries with thinness and extreme abstainers.

Just take a look at the image of Krishna, falling in love with the shepherdess Radharani, and you'll realize that firm and palatable flesh, and not puritan gauntness, is standard. Yashoda, Krishna's mother, is a woman with voluminous buttocks, borderline steatopygia, very feminine and sensual.

Returning to Buddha now. You can find him everywhere, from monasteries high up in the Tibetan ridges to little souvenir shops fully stocked with his image, manufactured in Singapore or El Salvador.

But if you wish to see him in different positions, all explained by Buddhist mythology, make an effort and travel to Thailand. From the basic Buddha, meditating in lotus position, to the reclined Buddha, comfortably and happily sleeping, or perhaps already departed from this tumultuous and stressful life.

You can find all of them there, more than forty positions, and showing off in all of them, with no false modesty, is his very voluminous belly and dewlap.

Is this perhaps his profoundest lesson to us, simple mortals without enlightenment?

Caviar

True caviar is eggs from the sturgeon fish.

There are other fish eggs in the market, some of them, like the mullet, can seem very similar, but they are not caviar.

Salmon eggs or red caviar are very delicious, but they're not caviar either.

Persians, back in the period of Cyrus, already consumed caviar, but in time, it became food for fishermen or for the very poor.

The Russians were the exception. They always liked this delicatessen.

Peter the Great, an avid caviar fan, sent, with his supreme ambassador, a box of caviar to Louis XV of France. King Louis tried the caviar in front of his court and immediately spit it out, thus creating, with his gesture of disgust, a diplomatic incident and an implied veto for its consumption by nobility gastronomes.

Caviar was not very popular in France, until Russian immigrants who fled the Revolution of October 1917, the Lenin Revolution, expanded their fondness for this food throughout Europe.

By then, King Louis' gesture was left in the past, and a new trend peeked in.

Very few people know that impoverished North Americans, at the beginning of the 20th Century, consumed large amounts of sturgeon eggs from the Delaware

River. Back then, the Delaware was the habitat of an enormous amount of sturgeons, and the wealthy hadn't yet discovered the gastronomic glamour of caviar.

Charles Ritz, owner of the Ritz hotels, turned caviar into the beckon of elegance and good taste when he ordered to include it in his hotel menus (starting in 1925), but beware, spittoons were placed very discretely near the tables... just in case...

Russian communists, very strict in almost everything, did not mind offering the monopoly of caviar to the Petrosyan family, counter-revolutionary immigrants but very good entrepreneurs; even today, they still maintain a good part of the multimillion dollar caviar business.

Ah! Adolfo Hitler, the German dictator, didn't like caviar either. He banned it from his personal table, alleging it was not worth it, and that it was too expensive.

Hitler, a maniac and a very strange man, asked the price of everything he and the people close to him were served.

His chef, very inspired once, tried to praise him with a beautifully served caviar dish, but since Hitler knew the price of that dish, he almost had the poor chef executed!

Or perhaps he deserved to be executed for praising such a terrible person?

Experts believe that the best caviar in the world is the Beluga, from Iran, which is the most expensive. If you save enough money to pay for it (it is quite expensive!), don't do what Louis XV did, maintain your poise and enjoy it; it's worth it!

CHAPTER 7

Rome. Keeping in shape
to conquer the world

Legend tells us that the city of Rome was founded over seven hills and at the banks of the Tiber River, by twins, Romulus and Remus, who had been raised and breastfed by a wolf. The legend also assures us that the year was 753 BCE, which has not been reliably proven but it must be close to the truth.

According to historians, Rome is an acronym of Romulus and Remus, but not everyone agrees with this.

What we do know is that Rome was founded by a Latin community, since prior to this, the Etruscans had already given life to this region (even though their origins is also topic of discussion).

Etruria, a very mysterious and interesting culture, occupied parts of what today would be Tuscany, Lazio and Umbria, where, among other remains, many burial sites have been found. They draw a lot of attention because of the size of the sarcophagus, many of them prepared to accommodate obese people.

The Etruscan civilization peaked between the 9th and 1st Centuries BCE, and extended towards Corsica, current Naples, and what later would become Rome, the same Rome, which with knowledge and prepotency would mark the end of the Etruscan.

The Archeological Museum of Florence holds the sarcophagus of Chiusi, an Etruscan, probably a nobleman, given its richness, and on it is imprinted the image of a very obese man, especially the abdomen, with a joyful and happy face that could be the prototype of happy obesity.

The Etruscan loved sweets and seasoned meats, and some researchers have posed the possibility that excess food and carnal pleasures had a lot to do with the decadence of this civilization.

The truth is that this attachment to the good life and the pleasures of flesh and good food, limited them in the development of their physical and military capacities, something that seems not to bother them much.

Catullus, who could not stand the Etruscans because of their lack of diligence and discipline, called them "obesus etruscus".

There must have been a reason for that.

For about 200 years, Rome was governed by a succession of kings, but in 509 BCE, it became an oligarchy republic, remaining this way until the accession of the empire in 27 BCE. When the empire is born, Rome was already a great world power, in the understanding that "world" refers to the basin of the Mediterranean Sea.

The family was the central nucleus of Roman order, the "gens", always headed by a man, the patrician. The woman, legally subordinated, maintained her submissive appearance, but always found a way to rule behind the throne. Some of these Roman matrons accumulated so much power that different noble families have gone on in history with the name of these female founders.

Roman art is not characterized by their originality. Its two basic sources were the Etruscan art, very colorful and surprisingly erotic, sometimes even pornographic, and the immense Greek art, which the Romans nurtured from, copied, and discretely ransacked, especially after conquering continental Greece and its adjacent territories.

Roman painting, very inspired by the Etruscan, is known thanks to a tragedy: the eruption of Vesuvius in the year 79 CE, which destroyed the cities of Pompeii and Herculean, but preserved, under ashes, a good part of their homes and installations.

There is no difference, except for the enormous talent of the Greeks, between the Hellenistic sculpture and Roman art.

Roman culture was permeated by the warrior and militarist sense of its founders, idealizing their image by representing them younger, stronger, more muscular, and fierce. Women are not represented as much, and when they are, they look quite similar to Greek women.

Romans were really huge in architecture, not only in the Italic Peninsula, but also in many of the territories they conquered and helped populate. In places as far as England and Egypt, we can still find today Roman buildings that survived the passage of time and the transformation of civilization.

Roman medicine was characterized also for being a continuation of the Greek (many Roman doctors were in reality Greek slaves). Galen, who would give his name to doctors around the world and for all generations to come, influenced western medicine until the advent of Renaissance.

But were there overweight people in Rome? Of course there were, and they could, without any complex at all, rise to very important social rankings. In his first 17 years of life, Lucius Domitius Ahenobarbus lived happily with that name, but suddenly, his life completely changed. The Pretorian guard, who had the responsibility of removing and putting in emperors, suddenly elevated him to the rank of Caesar while he was still an adolescent.

Faced with such an unexpected event, Lucius, following tradition, decided to change his name. He adopted the name of Nero Claudius Caesar Augustus Germanicus, although to his peers and for history, he would be known simply as Nero.

Nero was the son of Agrippina and Consul Domitius, but his father quickly disappeared from the scene, which allowed Agrippina to marry Emperor Claudius, who, coincidentally, was her uncle. Caligula, who was the emperor when Nero was born, had already been eliminated a long time ago in a rather expedite fashion.

Agrippina's next step, as Roman matron of arms, was to have Claudius adopt Nero as his own son. When Nero was 16 years old, one year before being an emperor, Agrippina forced him to marry Octavia, legitimate daughter of Claudius. Claudius, then, as a gift, named Nero, along with his son Brittanicus, co-heirs of the empire.

One false step.

A year later Claudius was murdered. Do you by any chance suspect Agrippina? You may be right. Subsequently, General Sextus Afrainus Burrus, chief of the Pretorian Guards and a very close friend of Agrippina, appointed Nero emperor, in substitution, of course, to the deceased Claudius.

At the age of 17, Nero had already reached the peak and Agrippina got away with her plan, but... and there's always a "but"... Nero, an apparently weak and effeminate chubby man, decided, without asking anyone, to put order, his order, his rules of the game, on everyone around him.

The first thing he did, just in case, was to kill Brittanicus, Claudius' legitimate son. Agrippina, very happy, rubbed her hands together thinking she already owned the empire, but Nero didn't quite see it that way.

Being very patient, he waited three years, and then... he also had Agrippina killed (it is said that he killed her himself, but this fact hasn't been confirmed). Since women weren't his forte, he also eliminated Octavia, his official wife and cousin.

Nero was already an orphan and also a widower. So nothing stopped him from dedicating all his time to his male lovers: Dioforus, Aulus Plautius, and Sporus, among others. But, wait a minute! There's also his lover Sabina

Poppaea, who he married, being very much in love with her. She becomes pregnant very soon, which doesn't look good for Sabina—a very flirty chubby gal, according to the bust that exists of her. Since Nero suspects the coming baby is not his, he savagely kicks her on the stomach, in a jealous rage, until she dies.

When Sabina Poppaea dies, Nero discovers, to his surprise, that his power cannot bring her back, which marked him forever, and he inconsolably cried for her death.

The story also tells about a Syrian slave, Actea, who seems to have maintained recurrent carnal relations with Nero. If this were the case, she must have been a very intelligent woman, or perhaps very lucky, because she survived.

At the end of his mandate, which had a couple of positive things, such as the establishment of the currency and the destruction of the Port of Ostia, he was accused of the Roman fire in the year 64 CE.

Perhaps he wasn't really the author of this tragedy, and the accusation may have been brought up by the defamation of Tacitus or Suetonius, his sworn enemy. In turn, to defend himself, he accused the Christians, a small and little known sect, which ended up joining this crazy circus.

Christians weren't very significant back then, but the officers of the Senators and the Emperor began to worry, very seriously, about the Emperor's madness and paranoia. The revolution began in Gallia and quickly spread throughout the empire. Galba, military governor of Hispania, marched with his legions in Rome. Nero carried out some desperate attempts to maintain control but he soon realized (he wasn't a fool) that he was too late to remedy the situation.

He decided to take revenge of the ingratitude of the Romans, banning them of their biggest artist: himself. He placed a dagger on the hands of his slave, and occasional lover, Epaphroditus, and ordered him to cut his throat.

Rome could once again breathe peacefully.

The phrase "bread and circus" will remain always as a paradigm of political pragmatism, which was enjoyed by Roman patricians and governors. In reality, besides the circus, the common people of Rome asked for bread, wine, and oil, with which their basic needs were met.

On an interesting note: Greek doctors made their careers in Rome, as we explained before, and one of them, a slave named Metrodorus, who worked as midwife and healer of women (a gynecologist) in the 1st Century CE, left us a good description of a type of anorexia nervosa that afflicted some young upper-class women.

Even back then!

Prince Esterhazy and his fabulous taste

Prince Pal Antal Esterhazy lived from 1776 to 1866, exactly ninety years, which back then wasn't a short time (nowadays you can say it's ok).

He was an outstanding member of the Austro-Hungarian nobility, close friend of kings and emperors and the Foreign Minister of the Empire.

But today, he is not known for his achievements in foreign affairs, but because of his "pie", of which he was, by the way, very proud of. The pie was prepared with five circular layers of compacted dough made with eggs, sugar, butter, grated almonds, and flour, separated by four layers of cream and decorated with lines of chocolate and chopped dried fruits.

Imagine! His coat of arms was printed on top of the pie.

But his culinary taste didn't end there.

He loved truffles (today you could say he was a fan of truffles) and he paid for them at the price of gold, and this can be literally said.

A truffle will never kill you, unless you die of fright when you see the bill. You won't gain weight from eating it either.

Nowadays, real truffles are the most expensive dish in the market.

One kilo of white truffles can start at $7,000, and the black ones, which are more common, are about $2,500.

They have a penetrating scent—like that of gas in a kitchen—and the taste is a bit bitter and unique, difficult to describe, especially, as is the case of the author of this book, if you have never tried it (I hope I do someday).

The descriptions are endless. They range from "celestial" and "fabulous" on one extreme, to "it tastes like carton" on the other extreme; but the latter, even though it comes from a renowned chef, may be taken a bit as contempt and envy.

And Prince Esterhazy? Well, if he were alive, he would be eating it, and not only his pie and truffles, but other delicatessen as well, and according to his contemporaries, in good quantities, too, and probably swallow down with wines from the Rhine.

How could he have lived to 90 eating like that?

CHAPTER 8

Blame it on the apple

The Bible, by definition, is a compilation of books, some of them accepted by all monotheist religions and others only by a few, and even others, known as apocryphal, by none. It is correct to clarify that we only know parts of some of these books, and others, which we do have only a reference of, have never emerged. The period of contemporary history that is covered by the books of the Bible includes almost 30 centuries.

In general, it appears that the Bible is just a book of history written by the Hebrews. The division of the Old and New Testament is Christian, and this latter compilation of books, much modern, is not recognized by other monotheist religions.

In this chapter, we will limit ourselves—this book deals with something very particular: obesity—to refer to certain characters and comments that may relate to its topic.

Genesis, the first recognized book of the Bible, is like an introduction or a prologue that justifies the existence of a Jewish community, which begins to be narrated in the second book: The Exodus.

Food is tightly related to Genesis. Man falls upon sin and loses paradise because of the evil serpent who offers Eve, the first woman, an apple.

Evil and disgrace of the human kind, which are translated into the obligation to work in order to survive and give birth with pain, began with the fruit. Today, we attack fats and defend fruits, but it seems that our ancestors didn't see it that way.

Adam and Eve's children, Cain and Abel, created agriculture and animal breeding. It was them, according to the Bible, who brought all the benefits of food, and we also owe them for all the nonsense of food, even fat.

That Cain killed Abel because of God's regards towards the quality of the offerings he received from them, all food, was already a premonition of what would come in the future.

The Exodus tells us the story of the manna that fell from the sky. Manna is a vegetable, a bit like cilantro, that has to be toasted and reduced to flour, with which a type of bread, very difficult to digest, is made. We could say that it is a poisonous gift. With this type of food, and walking all day, it was almost impossible to gain weight.

The Hebrews didn't have an extensive culinary culture, a phenomenon that has continued through time. Their land was difficult to harvest and they didn't experiment much with other varieties.

In regards to animals, the limitations imposed by religious guidelines were immense. The prohibition of ingesting pork may have had a practical reason, since trichinosis is transmitted through animal meat, and it had no cure back then.

Already in Leviticus, the third book of the Old Testament, sacrifices made to Yahew are already regulated, among other things. It is quite clear that God prefers the meat and blood of the lamb to any other animal, and vegetables and legumes are meant for mankind.

An offer that any vegetarian or vegan of today would approve.

Hebrews were very liberal in the consumption of wine. This can be confirmed by reading "Numbers", the fourth book. While tasting their wine, they could forget, for just a few minutes, the countless limitations and prohibitions.

Deuteronomy (Book of the Second Law) is, in a way, the story of Moses' farewell, and in Chapter XIV, verse 21 of this book we find a paragraph that has been cited many times as reference to a certain degree of wickedness and cruelty: "You shall not eat of anything that dies of itself: you shall give it unto the stranger that is in your gates, that he may eat it; or you may sell it unto a foreigner: for you are a holy people unto the Lord your God."

Nowadays, any respectable sanitary organization would close and fine a business that operates this way.

The New Testament, much more modern, as its name suggests, is not shared by the Jewish religion. It is the basic source of current Christianity and it is also made up of several books: 27 in total, written in the form of messages. It offers us more information about social and culinary traditions of the Hebrews and their issues are much more centered in the social problems of the community.

Reading the New Testament, you will be left with the impression that these people were, almost always, vegetarians: unleavened bread, dried fruits, dates, olives, onions, beans, cucumber, legumes, some cereals, lentil stew, and sometimes perhaps fish.

Hebrews were not familiar with sugar and their main caloric source was oil. Meat, especially lamb, was set aside for sacrifices carried out in temples.

It is possible that perhaps during Passover, after the priest sacrificed the animal, the family would take the meat home, and with an annual exception, they would eat it. Poultry would be eaten sometimes, but it was very expensive. Crickets and grasshoppers were common and a much appreciated source of protein.

In the Gospels, 27 miracles by Jesus of Nazareth are accounted for. Three of them have to do with food: 1—the miraculous fishing in the Sea of Galilee, 2—the conversion of water into wine during the wedding at Cana, and, 3—the multiplication of the bread and fish for the multitude of followers. The latter is the only one that seems to be narrated, in one way or another, in all Gospels.

Note that none of them have anything to do with cattle, pork, or poultry meat. In the Last Supper, one of the most famous evangelic events in history, Jesus toasts with bread and wine only.

Images in the New Testament portrait Jesus as an extremely thin man, almost undernourished, especially in the period very close, and during, his crucifixion. They reflect the kindness and enormous sacrifice he is making for humanity.

Caiaphas, the priest who condemns Him, is fat and has a big belly; he is a glutton and is synonym of injustice and evil. But you must also remember that these images were added much later from the actual time these events occurred, generated mainly by medieval and renaissance iconography.

Rome, back then, didn't take into account, except in the military, the wretched communities of Palestine and they, in turn, had barely any significant art, precisely due to the fact that they were underdeveloped and poor.

Gluttony

Gluttony is a vice and portraits bad manners. Also, for Christians, it is a capital sin, as described in the Bible.

Dante, in his Divine Comedy, condemned all the gluttons to stand between two trees loaded with fruits without being able to reach any of them. They would then go hungry for centuries to come.

Proverbs from the Old Testament are still even harsher, recommending the knife to detain evil. Terrible!

Gluttony means eating and drinking in excess, in an exaggerated and grotesque manner, without being concerned about keeping good table manners and forgetting about elegance and formal education.

In ancient times, drinking alcohol in excess and getting drunk, was considered a form of gluttony, but that has ceased (it seems getting drunk hasn't).

Gluttons seem to also disregard their personal appearance.

Furthermore, gluttons seem to also suffer metabolic ailments because of the large amounts of fats and sugars they consume.

CHAPTER 9

Hiding beauty at all cost

On December 1859, in her mansion in Paris, the wife of Prince Charles Louis Napoleon Bonaparte, the beautiful Eugenia de Montijo, was engaged in a heated argument with her husband. She just saw the paint that artist Jean Auguste Dominique Ingres—an old decrepit and salacious man—as she contended, offered as a gift to the young prince. He resisted for a moment, since he really liked the painting, but at the end he gave in to the vehemence and indignation of his woman and ordered the return of the canvas. The painting was barred from the palace, but was forever made part of history.

We're talking about an oil painting over a circular canvas, a little more than a meter in diameter (108 cms.), depicting, as the artist imagined it, about 25 young women, completely in the nude, and in different erotic positions, some of them even offering each other tender and lesbian caresses. The scene took place, supposedly, inside the great harem baths of a Turkish palace in a period, difficult to determine.

Years before, Ingres had read a letter from Lady Mary Wortley Montagu, dated April 1st, 1717, written to a friend, Elizabeth Rich. In this letter, written after a journey around Muslim lands, which caused great sensation at the time, there's a paragraph that obsessed Ingres for the rest of his life: "I think there were a total of about two hundred women; the first sofas were covered with pillows and extravagant carpets, where ladies were seating down; and in the second set of sofas, behind them, their slaves were seating. All of them in their natural condition, in other words and in plain English, portrayed exactly as they came into this world. Many of them were perfectly proportioned, like any goddess painted with the brushes of Tiziano. I loved their courtesy and beauty. Any man found in a place like this is sure to die."

Today, we can see a second version of this painting at the Louvre Museum, almost identical to the first, painted in 1863, when Ingres was already 83 years old.

The painting established Ingres as one of the greatest advocates of romantic realism in the arts, he was very educated, which is one of many reasons why he took this risk, for Picasso and other great painters, and besides, he gave the name to what we know today as "Turkish bath". It would also be a jubilant call for all plastic surgeons and lyposuctionists of the planet, since Ingres' taste for flesh, perfectly reflected in his painting, was similar to that of Rubens.

We've started this chapter talking about this scrumptious anecdote in order to illustrate a fact that has gotten much attention from historians throughout the centuries, and that is the almost absolute absence of the representation of the human body, especially the female body, in the Islamic art. The solid beauties that Ingres painted in the "Turkish baths" were only embedded in his mind—luxurious—without a doubt, but not in solid historic or documentary factors.

The practice of "jihad", the code that establishes the requirement of not showing parts of the female body in public (thus, the use of veils, tunics, cloth to cover their faces—nigab—full-body black cloak, and burkas, which only allows the woman to look through a piece of cloth), besides the absence of literary descriptions or even biographies, has made it impossible for the western civilization to know exactly what Islamic women look like under their garments, and I guess in a way for the men, too.

When the Greek philosophy was already discovered and discussed for two centuries by educated men, when the Roman Empire dominated enormous extensions of land on the east and west sides of the Mediterranean, and the Buddha had already been part of mythology and human history for a long time, Islamism didn't exist yet. Arabs, inhabitants of the Arabic Peninsula (currently Saudi Arabia and parts of Syria, Yemen, Kuwait, Iraq, and the Emirates) made up clans and tribes from the dessert, the Bedouins, who adored the stars and some stones, and in the coasts and to the more humid north, they had built some small towns, such as Palmira and Petra.

The coastal kingdom of Saba, where Yemen now stands— that of the beautiful Hollywood-looking queen who was allegedly Salomon's lover—had disappeared about a thousand years ago. And these Arabs were not yet Islamic.

Everything started with Muhammad, the last of the prophets, who came after Musa (Moses) and Isa (Jesus), but the most important and the last in the chain, according to the Muslims.

This man, who lived about 62 years (570-632 NE), never traveled outside the limits of the Arabic Peninsula, where he was born. He was a merchant until the age of 40, and from then, he became a preacher and sometimes a warrior, even though with unfit military fortune.

He left his followers a cluster of verses expressed verbally, since he was illiterate, so that they could put them together in the form of a book, the Coram, which would become the guide and source of the truth revealed to millions of people. Muhammad, a contradictory man in his actions—he first got married with a woman 20 years his senior and then with a 6 year-old—he had, without a doubt, great power of persuasion, proven by the fact, among others, that he was able to convince his followers that he traveled to Jerusalem in one night and in another to heaven to be interviewed by dead prophets, or when he justified, in the name of God, mass decapitations, such as that of the men in the tribe of Banu Qurayza.

One of Islam's most common practices is fasting. On different dates, throughout the year, but during the month of Ramadan, they must fast every day between sunrise and sunset.

It could be and should be a form of purifying the body; therefore, of weight loss, but what usually happens is that they consume exaggerated amounts of food during the hours they are allowed to eat.

Banning the consumption of alcohol could also be beneficial for their health, if it wasn't transgressed in so many different ways.

It is thanks to the Arabs that sugar is introduced in Europe.

They didn't discover it, but they did spread the harvest of sugar cane, and its subsequent industrial processing, even in Spanish lands, which would later allow for its arrival in the Americas. It is known that they really enjoyed homemade pastries and sweets.

In medicine, the Arabs also had their golden era. Between the 10th and 12th Centuries CE, when Europe was going through its darkest times, great scientific figures of Arab culture kept alive the flame of the Greek and Roman knowledge, incorporating many contributions of their own: Abdallah Ibn Sina, Avicenna for us (980-1037), wrote about the importance of being frugal when eating in order to avoid, among other ailments, obesity. Averroes, Maimonides, and other wise men also wrote, in the form of aphorism, about this issue.

The Renaissance culture in Europe owes a lot to the Arab world, unfortunately, nowadays, the cultural and scientific fizzle from that era is no longer observed.

Gastronomic plagiarism?

In 1879, the seal of arms of the Trinity College began to be printed, with a hot iron, over the flat surface of a cold cream dessert covered with a very hot layer of caramel.

Obviously, they named it "Trinity Cream".

This brings us to the conclusion that this delicious dessert, which later became known as "Crème Brûlée", was in reality English. However, the French adopted it and made it their own, making us all believe that it was born in France.

But things got complicated.

The French, very picky with their famous gastronomy, demonstrated that in Francois Massialot's cookbook, published in the year 1691, the instructions to prepare this sophisticated dessert were already detailed.

The case seemed to be resolved in favor of the French.

But no...

Massialot called his dessert... "Crème Anglaise"!

In reality, it's the same thing. It's a very delicious dessert and it is eaten today around the world.

Oh! And it's a dessert prone to plagiarism!

The Spanish have been calling it, for more than a century, "Catalan Cream".

CHAPTER 10

The culture of rice

Did Marco Polo really go to China? Because even though some of his contemporaries and researchers have doubted it, it seems that Marco Polo did in fact travel to China through the Silk Road. He spent about sixteen years in that strange land and returned to Venice, his hometown, by sea.

Back in those days, and even before, China and Asia, in general, were a mysterious and exotic world, closed for westerners. A world from where luscious fabrics, like silk, would occasionally arrive, along with exotic foods and a certain cereal, which in time would come to nurture half of the planet, as well as invasions of ferocious and cruel warriors, such as Attila. This cereal was "rice".

So let's go over Marco Polo's journey and get to know the history of rice a little more.

The origin of rice—scientifically known as Oryza Sativa, is controversial. For some researchers, the first traces of wild rice were found on the foothills of the Himalayan, some 7,000 years ago.

For others, these seeds came from the Ningbo Valley, in central China, around the same time, or even before (10,000 years ago), and then, some other researchers place them on the foothills of the lowlands of the current Thailand.

It has been proven that around 1,500 BCE, in the delta of the Niger River, in Africa, there was a type of rice known today as Oryza Glaberrima.

The history of rice is closely linked to that of Asian countries, and its harvest is for them, besides a vital need, a source of myth and religious allegories related to fertility.

Ever thought about the practice of throwing rice on the bride and groom at the end of their wedding ceremony and before, how can I put this... consuming their first intimate relation?

The history of China, as well as that of rice, goes back in time.

Usually, we try to put together a timeline through ruling houses and dynasties that begin with three mythological figures—the three Augustus—and continues with five emperors, whose existence are debated.

The most recognized is the so-called Yellow Emperor (around 3000 BCE); then the historic dynasties begin, as well as the first Xia, whose beginnings date back to the year 2100 BCE.

This succession of ruling houses is linked by four millenniums, with wards, coup d'états, political assassinations, rebellions and all types of cataclysms, until 1912, when Puyi, the last emperor of the Qing Dynasty is overthrown by the so-called Xinhai revolution.

Chinese art is marked by three philosophical-religious currents: Taoism, Confucianism, and Buddhism. The painting is elegant, light and barely uses human figures. The fat guy in Chinese art is the Buddha, the enlightened, even though the third emperor of the Ming Dynasty—Yongle—who changed the capital of Nanjing to Beijing in 1403, appears in a rare painting as an obese and sedentary man, but with a very astute gaze.

We would have to wait until 1949 to find ourselves with another fat man, Mao Ze Dong. One of his photos became an icon, extremely expensive, actually, by American pop artist Andy Warhol. But that's recent history.

Besides rice, the Chinese also consumed wheat and barley. In times of hardships, the millet, a plant that is very resistant to climactic changes, became one of the few sources of nutrition of the common people.

Even though meat, any meat, was considered a luxury, pork became nearly the only source of animal protein they were allowed to eat.

Besides rice, the Chinese left us tea and chopsticks.

As the story goes, chopsticks emerge from the very old habit of cutting meat and fish in very tiny portions, making the use of a knife unnecessary.

The fact that the gastronomic use of the chopsticks does not extend to neighboring countries (such as Japan, Korea and Thailand) until the 16th or 17th century CE is quite interesting.

The Mongols, a nomad community from Central Asia, quickly extended between 1206, year in which Ysugei Baghatur, known by his emperor name Gengis Khan, unified all the clans, and 1368, last year of the Mongol mandate over the Chinese people.

It was actually Kublai, the grandson of Gengis, the Mongol Emperor who got to meet Marco Polo, for whom, according to Marco, he completed a variety of diplomatic and commercial tasks.

The Mongol culture is fundamentally expressed in tales, quite often verbal, and a mythology that is not very extensive. Its painting is not very rich, either. A community dedicated to military conquest should not praise sedentary and overweight people, but when we see old paintings that represent emperors, we notice they are men with round faces and chubby figures.

This perhaps is explained by the Mongol phenotype. It seems true that Mongol horse riders, some of the best in the world, would put small bags of milk under their saddles, to turn them into cheese with the movement of the horse, so they could eat it for days.

Koreans form part of a community that is accustomed to living in the mountains, supporting a very harsh and unpredictable climate. Their staple food has always been rice and Kimchi, a vegetable that has a certain resemblance to lettuce, and is prepared with salt and radishes.

They were warriors and hardworking, and they never favored fat; however, in the few photographs found of the two dictators that governed with iron fists the northern region of the country in the last sixty years, Kim Il Sung and his son Kim Yong II, you will notice that they are two moderately obese men, just like Mao.

Things of comrades, defenders of the people.

Japan is a narrow-minded community that has fought, always and very ingeniously, against the lack of natural and food resources.

Its history, like that of China, is long and complicated. The denominated Jomon period, between approximately 5000 BCE and 200 BCE, is full of myths and the absence of historic documentation.

Despite being narrow-minded, the Japanese nurtured themselves in autochthonous cultural and gastronomic ways, but also taking in some from the Chinese, the Koreans, and other communities, continental as well as from the Pacific Islands.

Shintoism, a philosophy with religious factors, is a good example of this Japanese cultural syncretism.

Buddhism arrives in Japan around the 5th Century CE, and is accepted by many; however, it doesn't affect Japanese art significantly. Food, based on rice and fish (associated to sushi) begins to take its own shape and create very subtle gastronomic ideas.

Major differences between Japan and the rest of the Asian countries are established in the Tokugawa or Edo period, around the year 1600, when the Empire becomes completely isolated from the rest of the world.

These are the years of the Samurais, the Bushido code (which inspired the Kamikazes at the end of World War II) and absolute nutritional self-sufficiency.

In 1868, with the Meiji Emperors, the country opens up to western technology, especially military, but it stills keeps its insularity from extreme nationalism and chauvinism.

All this is splendidly reflected in their rituals: The beautiful tea ceremony, the institution of the geishas (brought and carried by western literature and film), the unrestricted adoration of the emperor (even though his power was more celestial than human), and the ritual suicide or Seppuku, allowed only for men, and where the intestine, that essential but undesirable element of life, is violently expelled from the body.

Rice is today a nutritional factor omnipresent in the planet.

Its protein and vitamin value is low and its caloric value is very high, but its ability to associate itself so nicely with other nutrients, turn it into a vital part of the diet of many communities.

A risotto in Italy, sushi in Japan, rice and chicken in Cuba, Chinese fried rice, a jambalaya in the deep south of the United States, a rice milk pudding in Spain are delights that easily make us forget about those extra pounds for a least a moment.

Enjoy!

An Asian eccentricity: Sumo

If you wish to become a great Sumo wrestler, you must submit yourself to an adequate training, and above all, gain weight. If you decide to do so, you are required to be sponsored by a Sekitori, former wrestler, deep connoisseur of the techniques and venerated master. If the Sekitori deems that you have the right conditions, then you will be named Rikishi, or an aspiring Sumo wrestler.

This is how it all begins.

The discipline, which is very strict, begins at 5:00 a.m. with a shintoist ritual. Strength, courage, and concentration are prayed for. You will need them. Physical trainings last for hours and ponder on dances and group choreographies, above all, handling the heavy bodies of other students without losing balance. They line up, pushing the person in front while the other resists, and this goes on for hours on end.

And what's for breakfast? Well, there is no breakfast. Any ingestion of food in the morning will be utterly banned.

Don't be surprised. There are two reasons for this severe measure of training: the most obvious one is to template the character and strengthen the ability to make a sacrifice, but the second reason is more important, metabolically speaking, "startling" the adipocytes, and predisposing them to store as much fat as possible.

Right after noon, it's time to abandon Dohyo, the ring where training and wrestling takes place. Chankonabe begins now, which is a ritual that must be executed thoroughly and at length. In silence, and very steadily, the wrestler will swallow, for several hours and methodically chewing, great quantities of different fishes, steak, pork, poultry, vegetables, rice, and legumes, everything boiled and cooked. And you'll really be surprised now: Everything complemented with litters of beer—not water—beer!

And when he can't swallow anymore, he'll just take a nap!

The immense stoutness of these athletes—two very opposing terms to be used in one sentence—can be so extreme, that they may need assistance with their anal hygiene after using the facilities.

Of course, the most famous, and best paid, can rely on their own geisha to take care of these duties, and we assume others too...

Sumo is an old tradition in China, India and Korea, but its documented history begins in Japan, more than two thousand years ago. The debate of whether Sumo is a martial art or a sport has weighed a lot in its inclusion in the Olympic Games. Today, it has extended to several countries and Hawaiian, Mongol, Russian and Bulgarian world championships have already been held.

The dark side of this old practice is the great toll it takes on the health of these athletes: Diabetes mellitus, arterial hypertension, chest angina, severe arthrosis and arthritis of the knees, cirrhosis, encephalic accidents, and sudden death, among other ailments.

The life expectancy of a Sumo wrestler is about 60 years, almost 20 years below that of the present average Japanese male.

CHAPTER 11

The land of cocoa

Any philatelist can show you a 32-cent U.S. stamp with the portrait of a gray-haired man with a mustache, and a very kind face and smile. Around the seal it reads: Milton S. Hershey. Philanthropist.

This tells us something, but not enough about this man, also known as "The Chocolate King". Greg Rothman, Hershey's biographer, wrote that he was to chocolate what Henry Ford was to cars.

Hershey was born in a small country-side town in Pennsylvania in 1857. He came from very poor parents who practiced the Mennonite religion. He dropped out of school in fourth grade to work full-time on manual labor and he never went back to school. He was a tailor apprentice, but at the age of 18 he asked his aunt Mattie for 150 dollars to open his first, and very small, candy factory.

He worked all night making candies and he sold them during the day, but due to the high price of sugar in those days, his business wasn't very profitable. Seven years later, he decided to quit.

Of course, he didn't surrender for long.

He worked in Denver, Chicago, and New Orleans, always dreaming about returning to his own business. He tried making cough medicine in New York, but failed again.

At the age of 29, Hershey again asked for a loan and began manufacturing "Crystal A Candies", which were his own original candies, but now with chocolate. By the time his friends and families started advising him, in good faith, to quit his dream and look for a steady job, success unexpectedly knocked on his door.

He later explained that even in his darkest moments, he never left behind the quality of his products... and chocolate worked its magic.

In 1898, he married Elizabeth "Kitty" Sweeney, a girl from a Catholic Irish family who was sixteen years his junior. Since they couldn't have children, they both put all their energy to work, making chocolate, of course, and money. But even then, she had already established that they should put something aside for those less fortunate.

In 1918, at the age of 61, Hershey's fortune already reached 60 million dollars. He put everything in the construction of a city (Hershey, Pennsylvania), a university, a school for children with special needs, recreational centers and residences for the employees of his immense chocolate factory.

Kitty passed away at the age of 42 from a neurological disease. Hershey, who never remarried, took refuge in his work, his routine, and his solitude.

But one day he discovered the beauty of Havana, Cuba. He purchased an apartment there and began to negotiate with wholesale sugar. He never forgot that he once failed because of the price of sugar, and this motivated him to venture in this, now quite lucrative, business.

The next step was the construction of a modern sugar Factory outside the town of Santa Cruz del Norte, relatively close to Havana. He called it Central Hershey, and facilitated it with top-rated technology of the time and with U.S. conveniences for his employees: proper homes, running water, electricity, paved streets, a school, and a hospital.

The author of this book had the chance to travel once from the port of Havana to Central Hershey in the railway that was also constructed by Hershey to transport sugar from his warehouses to the ships anchored in the port, and in turn, facilitate his workers and the public in general the path through this area. A path that ran about 40 miles through a plain and highly beautiful territory.

It seems almost impossible to believe that after all the time that has gone by and all the political and economic disasters this country has suffered, this railway continues to operate today, even under its original name: "Hershey Railway".

But... where does chocolate come from?

When Christopher Columbus arrived to the New World, in 1492, nobody in Europe, Asia, or Africa knew about chocolate, and of course, he had no idea that one of the most famous and most enduring contributions of that land that he was just starting to discover for the Old World, would be precisely, the cocoa, in other words: chocolate.

Cocoa, fruit from which chocolate is extracted, seems to have originated in the basin of the Orinoco and Amazon rivers about 3,500 years ago.

Chocolate, as a beverage, a bit different from what we are used to drinking today, was made popular by the Olmecs, a Mesoamerican culture we know very little of and that we recognize because of their enormous and heavy stone sculptures of faces with flat noses, obese, and large.

From the Olmecs, the cocoa passed on to the Mayans, who even appointed a god for the fruit: Ek Chuah. Many years later, the Aztecs elevated it to a noble product, a beverage for emperors, and also to a more ordinary, but significant, category: money.

We consume the chocolate that is made from a solid paste and butter (fat), mixed with sugar.

The Aztecs, who were not familiar with sugar, drank it with spices, turning it into an energy drink, but not as delicious. This is how the European started drinking it, using it as a stimulating and aphrodisiac substance, but not as the delicious beverage we know today.

We're not quite clear who the first person was to mix it with sugar and prepare it in the way we consume it today. There are several accounts, but among the most extensive are those that attribute it (and thank) the cloistered nuns of the Guajaca Convent, in Mexico, and/ or the Monastery of Stone, in Zaragoza, Spain.

The Jesuits, a very powerful religious order in those days, had a lot to do with the fast dissemination of the use of chocolate in Europe and even in America.

The English were the first, as late as in the 17th Century, to mix chocolate with milk. Before that, it was mixed with water.

Back in those days, chocolate, with water or milk, was considered by almost everyone a medicine, even though gluttons were starting to consider it as gourmet. Recent neurobiological researches associate chocolate with some brain neurotransmitters and attribute it with anti-depressive effects.

In other words, it somehow goes back to being a medication.

The first U.S. chocolate factory was founded in 1755. Hershey came later. Bonbons were invented by the Italians around 1830.

Solid milk chocolate was invented by Daniel Peter in 1875. Kohler, Nestle, Lindt, Tobler, Ferrero and many others improved the presentation and developed the industry.

In 1912, the U.S. company Nabisco released the Oreo cookies. The Snickers, chocolate bar with peanuts, was invented by Frank C. Mars in 1930. In 1941, Forrest Mars (the son of the guy from Snickers) and Bruce Murrie introduced the M&Ms (Mars & Murrie), manufacturing today 400,000,000 a day.

That's enough! I'm taking a break to sip a delicious cup of hot chocolate.

Hunger, appetite, and satisfaction

Hunger is a sensation, measured by a part of the brain called hypothalamus, which occurs when our body needs nutrients and energy, and "asks" us for food, using this sensation as an alarm signal.

Even though you may find it difficult to believe, scientists cannot agree on this definition I just gave you.

Some suggest that more than nutrients, hunger indicates the need to have some volume of food in our stomachs, and others, not many, suggest that hunger refers to the lack of glucose circulating in our bloodstream.

Modern researchers have found several substances segregated by the stomach and the biliary vesicle, which, when transported through the blood, act on the brain, closely related to the sensation of hunger.

Appetite is easier to define than hunger.

Appetite is the ability to wish for food that we like. Appetite requires that a person be healthy, that his or her digestive and nervous systems be in good conditions, and it even has to do with gastronomic and cultural education.

When you are sick, you usually have no appetite. A person suffering from a brain injury may also not have an appetite.

If you are in an adverse environment, where the food being served may not be something you are familiar with or you perhaps don't like it, you may have no appetite, even though you may be very hungry.

A person lost at sea, with only salt water and perhaps a raw fish to eat, may be extremely hungry, but have no appetite.

When we're hungry, we can eat almost anything; but when we have appetite, we seek for food we really enjoy.

When you're healthy and feel no hunger or appetite, you may be satisfied.

In other words, satisfaction is a perception of our brain, also measured by the neuronal region of the hypothalamus, which stops us from continuing to want more food.

When we say: "I'm full!" What we're trying to say is "I'm satisfied".

Satisfaction also requires a healthy digestive system, an integral nervous system, and good health, since the lack of appetite presented by a sick or unconscious person is not satisfaction.

Hunger, appetite, and satisfaction are physiological sensations that are even submitted to several, and very interesting, scientific investigations that can present us with many surprises, even in the field of weight control and obesity.

CHAPTER 12

Byzantium,
a bridge between two worlds

The morning of May 28, 1453 was clear and quiet. For the first time in months, the constant thundering of enormous ottoman bombs were not heard, and mutilated bodies did not fall anymore upon the city walls. Sultan Mehmed consulted once again with his astrologists and they confirmed to him that the following day would be very unfortunate for the Christians.

Inside Constantinople, now almost in ruins, Emperor Constantine XI Palaiologos, understands perfectly well the menace of the enemy's silence.

He attends mass at the still splendorous cathedral of Saint Sofia; he kneels down and once again asks God to give him strength and courage to comply with his promise to die without taking a step back or pleading for clemency.

He stands up, and with proud voice he orders all the bells in the city to toll incessantly.

He figures it's better to daze the mind with the outcry of the bronze bells than silently ponder over the approaching infallible torment. He can only rely on what is left of his army; he encourages them with his gesture and leaves with his soldiers and knights to the barricades.

The day and long night go by, weighing in their hearts. An hour before dawn, on May 29, the vicious attack of the Turks unleashes. Wave after wave of warriors are thrown in ditches and walls, but the exhaustion of the defenders and the infinite number of attackers finally prevail.

As vowed by Constantine, he dies as a hero, or perhaps a saint, or a fool, according to Mehmed, who offered him wealth and praise if he relinquished his city without a fight.

Constantinople kneels down and gives up. The Christian world, which didn't do much to resist, is horrified. It wasn't just a city or an emperor that plunged, it was an entire world. A bridge between two nations was smashed.

The Middle Ages was over, drowned in fire and blood, even though nobody at that sinister time realized this transformation.

The Byzantines are Greeks; Byzantium was the capital of Thracia, a territory that was traditionally Greek, where the mythic, and not so mythic, Troy was founded in ancient times. Its history was linked to the turbulent centuries of Peloponnesian wars, the rivalry between Athenians and Spartans, the battles against Cyrus, Darius and Artajerjes, who preserved the Hellenistic culture and the Macedonian period, led by Alexander the Great.

With the centuries, Byzantium became, by order of Emperor Vespasian (9-79 CE), part of the Roman Thracia. In 330 CE, Constantine I, the Great, changed its name to Constantinople, and gave it sort of a category of subsidiary of the Roman Empire in the east.

Its location, at the entrance of the Bosphorus Bridge makes it strategically essential as a link between west and east. "Byzantine Empire" is a term established by renaissance historians. For the "Byzantines", they were just Romans from the east region.

Byzantine art, fundamentally religious, constitutes a fusion between classic Greek art and the primitive Christian iconography.

Unfortunately, the preponderance of the so-called "Iconoclasts" for an extended period of time, limited the development of sculpture and paint with human forms. Architecture is the great byzantine art. Sculpturing is poor, and painting, almost always in small format, is quite instituted by Paleochristian imagery.

The Byzantines established their very own eating etiquette. They eliminated the uncomfortable habit of lying down to eat, which was a Roman custom, and now sat to eat.

It is reported that they invented the fork. In good measure, they created what is now called the Mediterranean diet, or at least they contributed to it.

They consumed great quantities of yogurt, prepared salads with fresh vegetables, seasoned with olive oil and vinegar. The "sadziki", for example, is a cucumber salad cut into cubes, seasoned with olive oil, garlic, dill, vinegar, and yogurt, mixed together until a paste is formed, also used as a bread spread.

Pilaf rice, aromatized with clove, bay leave, raisins, cardamom, onions, and almonds, typical of Turkish cuisine, is byzantine.

Their gastronomy closely followed the principles of Hippocratic medicine, which was healthy, and in a way, very modern. They ate very little red meat and didn't view it as a healthy choice.

They considered that food had to be fresh, full of natural elements, and harvested close to where they would be consumed.

Looking at them with today's vision, we could say that the Byzantines were very "organic".

The changing vision of child obesity

Max Fleischer was a genius of animation.

In 1921, he created, together with his brother, Dave, and a small group of enthusiastic cartoonists, the Fleischer Studios, which would in the next ten years give life to the most famous and charismatic characters, such as Popeye, Bimbo, and Superman.

In 1930, they demonstrated their immense talent with a figure that would make, and still makes, history: Betty Boop, the flirtatious and slender girl that almost shut down his business because of the Hays Code, who censured and accused them of "sexual blatancy".

But the Fleischer's couldn't keep up with the great film studios and were swept away off the market by Paramount.

Warner Brothers, ferocious rival of Paramount, then hired one of the men who worked for Fleischer, Leon Schlesinger, gifted with a great ability to coordinate teams.

And so, the Looney Tunes branch was born, whose passionate history is intimately linked to the emergence of sound films.

From the Looney Tunes workshop, a gold mine, we will limit ourselves to remembering one of those unforgettable characters, was officially born from the brush of Friz Freleng, one of Schlesinger's subordinates, on March 2, 1935: Porky Pig, a little fat, asexual and stuttering pig, who started as a kid and then grew up. He even had an ambiguous and also asexual partner, a bit like a sister, a bit like a girlfriend: Petunia.

In a time with little social complexes and where obesity was not viewed as an epidemic problem, even less as a way to put down children, Porky Pig was a huge commercial success, starring in 152 short films and some feature films, even conquering television with its very own show, besides circulating around the world, in pages of newspapers, magazines, and comics.

A great commercial deed, but times have changed.

In 1991, the sky grew dark for Porky Pig.

The National Stuttering Project (NSP) of San Francisco filed suit against Warner Brothers, in support of a New York man who was cruelly abused by his school peers because of his stutter and obesity.

For years, wherever he went, he endured being called Porky Pig. They settled for $12,000, but nothing would ever be the same.

Children obesity began to be viewed as what it is today, a global epidemic that continues to grow. All those jolly fat characters: Dumbo, Piggy, Petunia, Winnie the Pooh, Yogi Bear, Barney, and many others began to decline, their golden age had been left behind and the future did not seem gleam at all.

In 2001, a team of inter-disciplinarians of the Japanese company Nintendo, directed by Shigeru Miyamoto, released the first Wii system, with the idea that children should interact with each other and continue moving.

Every year, this technological accessory becomes more complex and efficient. Children, as well as adults, can now actively participate in sports, for hours on, in a small space in front of their TV.

How distant and archaic now seems that poor and almost forgotten chubby Porky Pig and his cheerful and fleshy girlfriend Petunia.

CHAPTER 13

Monotony, misery, and gluttony.
The Middle Ages

A medieval knight is welcomed inside a castle near Avignon. He has been traveling by land from Salerno—port located in the Italian Peninsula—to his home, in the Montpellier region. Weeks had gone by since he disembarked in the deafening dock, from a galley originating in the Holy Land, and only a few times was he able to sleep under cover. He also hadn't eaten a hot meal, and even though his suit of armor, spear, and shield was being carried by a scrawny esquire in a donkey, which thankfully he was able to purchase at the price of gold, his joints and muscles are still in pain.

This castle and its generous Castilian were put in front of him by our good God. The squire, his two horses and the donkey will be fed and sheltered in the stables, warm and dry, while he, the knight who proudly wore the Holy Cross over his chest on his tunic, will be conveniently welcomed by the proprietor and his servants.

By sundown, everyone moves to the great stone hall of the fort; the cold air starts to creep in through the cracks, but some beams and coarse skins prevent it from directly hitting the back of the guests.

They also warm up with the crackling fireplace, continuously attended to by a lad. The smoke makes them cough, and their eyes are itchy, but they pretend everything is fine, since it's always been this way, and besides, this is much better to be out in the open field.

The banquet begins with warm red wine, a bit vinegary, but it is harvested at home and that is always appreciated.

Then they bring in the entire deer, decapitated in front of them; it is fresh, since it was hunted specifically to welcome him, and just for plain satisfaction and delight, they season it with a little salt. The bread is scarce and hard, but when damped in wine and the gushing fat, it tastes like heaven. While cleaning his hands in the fur of a Great Dane that lies by his feet, the knight sees, up in the shadow, the lady of the castle.

She greets him from afar and moves on to the kitchen. He can't quite tell what she looks like, he hasn't even been able to really see her face or her eyes, he doesn't know her age, or whether she is fat or thin, or whether she's pregnant, which is very common in those solitary locations, but none of that is important.

The heart of the knight jumps and his hands begin to sweat. He already loves her and his life has forever changed.

He will pay a servant with Saracen money, to steal a piece of cloth she lady may have touched with her hands, perhaps a dirty kitchen towel or a thread from her slip, which from now on will always be tied to his shield.

He will never again see her or know what became of her. Whether dead or alive, something he will never try to find out, she will be his lady and his light on this earth.

At dawn, the knight kindly thanks the Castilian, the gentleman who nobly welcomed him, and after accepting a few slices of old bread, he continues on his way, solitary and cold.

Now the man is complete; he has fought the Holy Sepulcher, has traveled, has suffered and has a maiden to whom he will always dedicate his victories and for whom he shall die, if necessary. She will never be his carnally, that's what mistresses are for, but she will always be present in his tales and songs. Ah! And he must make up a name for her according to her dowries.

This story, for us a bit absurd, was very common for the reduced number of nobles and feudal gentlemen who did whatever they wanted during the Middle Ages.

Centuries later, Miguel de Cervantes wrote the most paradigmatic book in Spanish: "The Ingenious Hidalgo Don Quixote de la Mancha", precisely to make fun of things like these.

Saint Catalina de Siena, a cloistered nun in a medieval convent, disciplined herself with a steal chain to keep away evil thoughts, among them those related with the desire to eat. She lived off bread and water until the day she died.

Historian Rudolph Bell called this masochistic conduct "Holy Anorexia".

Sin resided in the body of women, source of all temptations, which had to be martyred to turn it into something androgynous and not quite attractive, which makes us think that medieval men in reality liked chubby women.

Joanne of Arc, another character who was continuously fasting, passed as a man, a boy actually, when she dressed in her suit of armor and carried her sword.

Saint Augustine (354-430), at the very beginning of the Middle Ages, made it very clear that beauty, whether in real life or in art, derives from the Lord's concepts, making inner beauty the only important matter. A Christ is beautiful if people wish to adore Him in Church.

On the other side of the coin, Friar Tuck, overweight, strong, and charming, loyal adventure companion of Robin Hood, is the image of the rebel who confronts the feudal, despotic and abusive warlord, and to hurt and benefit himself at the same time, he hunts and eats, until gorged, forest deer that are the absolute property of the lazy gentleman, who prevents his subjects and vassals access to a decent meal.

An interesting and very illustrative anecdote, and unlike those of Friar Tuck, historically very well documented, is that of Sancho I "The Fat" (935-966), King of Leon, in the Spanish borders with Castile and Navarre. This young man, instead of dedicating himself to politics and war, tasks associated to his royal status, spent all his time eating, which leads him to become a true barrel!

His obesity, coupled with a lack of intelligence, according to the chronicles of the time, incited in several gentlemen, the desire to oust him, which they succeeded in the year 958. Sancho sought refuge in the colonies of Queen Toda of Navarre, who was his grandmother.

This lady, who wanted to recover her privileges in the Kingdom of Leon, submitted poor Sancho to a rigorous diet, one of the first in record, which included sewing his mouth, leaving just a hole, through which he could slurp liquids. He was nourished, seven times a day, with salt water, lemon balm, honey, juniper, dandelion greens, and some vegetable concoctions. He was tied up to his bed to receive deep massages and was subjected to steam baths.

The treatment, which nearly killed him, gave satisfactory results in 40 days. Sancho, who was also compelled to walk, while being tied up, lost 70 Pamplona pounds, which we don't know exactly what it would be equivalent to, but we do know it was half of his initial weight.

With the help of the Muslims, Sancho regained his thrown in 960, and ruled, with increasingly more mistakes, until 966, when he was murdered with a poisoned apple.

A nice and appropriate ending to an insatiable eater.

Cookbooks. A Medieval invention?

If obesity and the lack of culinary taste were to completely dominate the world, cookbooks would become just like Manuscripts from the Dead Sea, in this case, archeological refuges of a past full of flavors and good taste, lost forever.

Recipes, transferred by word of mouth, are prehistoric and precede scriptures by tens of thousands of years. Gathering recipes took time, at least until the advent of writing.

We don't have an accurate account of Greek or Roman cookbooks; perhaps they didn't invent them; but if we take into account that they created philosophy, history, theater, poetry, quibbles, law and many other things, it's hard not to believe it.

Let's remember Socrates, who before drinking the hemlock, reminded his students they had to pay the rooster that was owed to someone. Perhaps to cook a soup?

The first documented cookbook is "De re coquinaria" ("On the Subject of Cooking"), written by someone named Apicius, around 400 CE.

Then we skip to the "Liber de Coquina" (Latin: The Book of Cookery), in the 13th Century, put together by French authors. And the first book of which we have copies of is "The Forme of Cury", by master chef Richard II of England. But at the same time, emerged—among the nobles, of course—"Viander", by the master, not only of the kitchen but also of good manners, Guillaume Tirel, also known as Taillevent.

By the 17th Century, cookbooks multiplied like mushrooms. "The Closet of the Eminently Learned Sir Kenelme Digbie Knight Opened", dated 1669, and in 1742, the North Americans (they weren't yet) came up with "The Compleat Housewife", written by Eliza Smith, who curiously was one of the first female writers of this side of the Atlantic.

A contemporary of Napoleon Bonapart, Jean Anthelme Brillat-Savarin, French doctor, attorney, and politician, made news with his "Physiology of Taste", a confusing and convoluted book (cited by almost everyone) that is saved by three truths, as valid today as they were in their time: 1—Those who eat fish often live longer, 2—Diets with sugar are harmful, and 3—Tell me what you eat and I'll tell you who you are, a judgment made famous once again by Chairman Kaga in the TV show Iron Chef.

Then came Auguste Escoffier, with more than twenty books between 1903 and 1934, who created a style and a form of doing things that still endures today, almost to the tee. By then, cookbooks became literature, and business, of course.

Describing current cookbooks is almost impossible: ethnic, economical, vegetarian, fast, colored, artistic, by artists, gourmet, for winners, for the depressed, as a joke, serious, literary (reading Isabel Allende), seafood, seaweed, religious, for athletes, and so, thousands and thousands of different books.

There are also oddities and "daunting" recipes.

If you're still in doubt, look for "The Alice B. Toklas Cookbook", published in 1954 by Ernest Hemingway's good friend and love companion, novelist Gertrude Stein, who, very seriously, details the "hashish fungi", a dessert, apparently very delicious, made with seasonal fruits, nuts, spices, and marijuana.

Note: Serve warm.

CHAPTER 14

Living, creating, and eating in abundance. The Renaissance

If Mrs. Isabella Brant and Mrs. Helena Fourment, specially the latter, lived today, they would surely be a pair of very kind chubby ladies, very pleasant, with hidden complexes and remorse because of their plump figures.

But to their good fortune, and that of art, they lived in another time, and they were both the greatest love of a gentleman named Pedro Pablo Rubens.

Besides a genius with paint, Rubens was a true aristocrat and an accomplished traveler and diplomat; he had no complexes (he had no reason to have them) and he loved his chubby women, so much that he had three children with Isabella and five with Helene.

Between 1630 and 1640, year when the painter dies at the age of 63, Helene was the delicious Venus in "The Feast of Venus", which we can admire today in Vienna, she was the plump beauty that Prince Paris must happily inspect in "The Judgment of Paris" (The Prado Museum), she posed for one, or perhaps more of "The Three Graces", also in the Prado Museum, she is also the sensual Eve in "The Fall of Man" (Prado), and among many others, the most famous "Portrait of Helene Fourment", also known as "Het Pelsken", in which the painter completely undressed her, even though he wickedly allowed her to cover her breasts with her right arm and her buttocks with a furry skin.

This painting, which the master kept to himself for a long time (evidently videos were not invented yet) can be enjoyed at the Kunsthistorisches Museum in Vienna.

When Rubens painted this marvelous piece, he had just met Helene, and he painted her without retouching it or trying to minimize her incipient fat. He paid tribute to her grace and beauty, evidently proud that she was his woman, without hiding her body nor her name.

He was a free man, a great artist, and he passionately loved Helene. A good lesson for us, right?

Like every artist, many influences are attributed to Ruben in his form of painting, among them, the classic Greek and Roman sculpture, and closer to our time, the Venetians Giorgione, Tiziano Vecellio, another sybarite lover of rounded female forms, the Veronese and Tintoretto, and of course, masters Leonardo Da Vinci, Rafael Sanzio and Miguel Angel Buonarroti. Good gracious! Wish all "bad" influences were like these!

And these influences, and some other ways of making things or thinking, were known later as "The Renaissance.

The Renaissance, or the Renaissances, was a cultural and scientific process that benefited European countries during different time periods and with diverse characteristics. They don't have an exact beginning, clearly defined with a date or historic event.

This period is about a progressive reunion with the ancient Hellenistic culture, a growth of scientific research and above all, of technical development: navigation, astronomy, press, anatomy, artillery, architecture, optical, cartography, etc., a liberation of habits, a greater access to information, the discovery of new worlds and customs and an awakening, almost explosive, of art, specially pictorial and sculptural.

Painting acquired death and color. Beauty ceased to be just religious to acquire its own value.

Extreme thinness continued being synonym of suffering and surrendering to God, but robust figures, especially female ones, symbolized maternity, even Christian maternity (like the innumerable Virgin Maries that populate museums and private collections around the world) and also enjoying life, the pleasure of a fine table, carnal love, and pagan eroticism.

The paradigm of Renaissance and the Renaissance man is Italian—even though Italy didn't exist as such—Leonardo Da Vinci (1452-1519), bastard son, born in Tuscany, one of the most beautiful sites of the Italian Peninsula.

At the age of 24, he is detained and taken to court for having had homosexual relations with a 17 year-old boy, charge that is dismissed, probably taken care of by his father, but which most likely may have been true.

At the age of 30, he is admitted as engineer and painter at the service of Ludovico Sforza, and his fame begins to expand: anatomical studies, armaments, fortifications, hydraulic projects, flying machines, canons and stone launchers, fireworks, decorations, observations over the animal and human movement and an infinite number of other masterpieces which he became interested in for a little while, and then many times, he left them incomplete.

His famous drawing "the Universal Man", also known as "The Vitruvian Man" is classical of the harmonic and perfect human measurements and a very clear sign that beauty is in the middle ground.

He was a brilliant man, a bit strange, probably overwhelmed because of his unpleasant experience in youth, which turned him into a distant and cautious man with his personal relationships, but he never lost his grace and courtesy. In 1516, he left to France as the official painter of Frances I, but he was already sick—he had a stroke that left the right side of his body paralyzed, which did not limit his job completely, since he was left-handed, but he was physically and psychologically affected. He died peacefully and resting, as the story goes, in 1519.

There's no doubt that the concept of beauty from the Renaissance period was inspired in the classic Greek and Roman imposition, but it also meant a return to recognizing the real form of people and things.

A beautiful model could be painted, but at the same time, she was painted exactly as she was in real life, and her life in those days did not contemplate extenuating diets, reconstructive surgeries or exercise at the gym; actually, good food, physical deformities and maternity were very important part of life and they were not rejected or hidden.

An example of the real expression of life in painting is the portrait of the obese English King, Henry VIII, made by Hans Holbein The Younger. This King, renowned because of his six famous women, ended his premature life because of his orgies and excesses, of which syphilis was a byproduct.

The dwarf "Morgante", obese and deformed, was the model in the fountain at the Boboli gardens, in the Pitti Palace, in Florence, executed by Settignano, and he modeled again for the work of Cioli and Giambologna.

Another aspect, one more, which Renaissance substantially and positively changed, was gastronomy.

The splendor and production of Renaissance banquets, far away from the griminess and coarseness of medieval times, was accompanied by an evolution in the preparation of new foods, their quality and freshness, the plating, the ornaments, a visceral rejection to monotony and boredom.

Even Leonardo specialized in these tasks.

Table manners also became very appreciated, which made elegance and good manners part of culture and nobility.

There was still a lot to be learned, but without a doubt, they were on the right path.

Giuseppe Arcimboldo and the food pyramids

Some modern critics have insinuated that he was crazy, even though they use the term "perturbed", which is less harsh. Most art connoisseurs have always visualize him with a curious personality and a second-rated painter, but nothing comparable to the great Renaissance figures.

Giuseppe Arcimboldo was born in Milan, Italy, in 1527. Son of a painter, and a painter himself, he spent many years traveling from city to city: Monza, Como, Cremona, Florence, Rome, Vienna, and Prague, where he worked for the court of the Habsburgs for 20 years.

Very sick and prematurely old, he returned to his hometown, Milan, to die in 1593.

During his long stay in Prague, he worked for three kings: the Emperors of the House of Habsburg, Maximilian II and Rudolph II, and Prince August of Saxony, who hired him in Vienna around 1570.

For the people in Prague, Arcimboldo was sort of a Leonardo Da Vinci, perhaps in a lesser scale.

He designed rooms, furniture, stained glass, tapestries, masks, choreography, and costumes for parties and celebrations.

He was a magnificent luthier and some of the drawings of the musical instruments he invented are still preserved. He also executed some interesting engineering projects, especially hydraulic, but they never went beyond mere curiosities without precise use.

However, his job was to paint, and as it was common in those days, he evolved in two main fields, religious and portrait art. With the first, his art was quite mediocre and almost fallen in obscurity, but the second has made its way in the history of art, where he is appreciated more and more each day. Let's get a bit closer to these strange paintings.

We are dealing with heads and faces, some specifically historic characters, such as Rudolph II, his patrons, a Vertumnus (Roman god of vegetation), a librarian from the court or a shyster, friend of the painter (an attorney), images within images, that change according to the position or angle it is looked at, or representations of the different seasons of the year, and elements of nature adopted human forms: spring, summer, fall, and winter, and water, land, air, and fire.

But the most fascinating thing about these masterpieces, and others painted by him, is the happy use of flowers, herbs, and other elements of nature that give form to figures, but above all, of the vast array of edible vegetables, fruits, cereals, fish, and on some occasions, other white meats, especially poultry, prominently in any food pyramid, so popular nowadays.

All these peaches and apples, strawberries, pumpkins, grapes, fresh asparagus, cucumbers, peppers, blueberries and a variety of cereals, lettuce and cabbage, garlic and onions, flounders and sardines, would delight the best nutritionists and the health of many of those who now pay the price for the excess fast food and canned sodas they consume.

Perhaps Arcimboldo wasn't a sublime genius of paint, but leaving his paint aside, he did contribute to the advances of the fight against obesity.

Let's all follow his advices, so colorfully administered.

CHAPTER 15

Slavery and sugar

Sugarcane is originally from India and southern China. People have known about it for thousands of years, but until Alexander the Great invaded Persia, Europe wasn't aware of its existence. Persians described it as "a plant that produces honey without the intervention of bees", which seems to be an accurate definition.

After Alexander, the harvest of sugarcane started to gradually expand, and its fundamental product, sugar, turned into a delicatessen for the palates, payable in gold. The Arabs traded with sugar and there is evidence that they had at least one factory in the south of Mozarabic Spain.

In 1498, in his third trip to Spain, Columbus brought to the New World—to Santo Domingo, specifically—samples of the plant, which perfectly adapted to the region and quickly expanded to other Caribbean islands.

Hernan Cortes took it to Mexico during the conquest of this nation and other conquerors introduced it to Peru. But its natural habitat, measured by its productivity of sugar, was the Caribbean.

And the Spaniards hated to harvest, besides, there were very few of them, so it was indispensable to get more manpower, preferably cheap, even better if it was free...

Caribbean natives weren't robust enough to perform the harsh work of cutting the cane, carrying it to the mill, and grinding it. However, they were forcibly used for this task at the beginning of the conquest. They were punished, mistreated, and starved, which wipe them out in just a short time. Stories of mass suicide of the Caribbean native, who preferred a quick death over the infinite exhaustion of this slaving work, are horrifying.

This is where we come to the turning point that led to the curse suffered by Africans and the Americas for three centuries: black slavery.

As the story story goes, Father Las Casas, in an effort to protect the natives, advocated the use of black Africans to work the sugar.

There's no question that this happened, and we have the letters and journals of the priest to prove it. But due to the shortage of native manpower, they would have brought in Africans anyway due to the simple reason that slavery already existed for some time now.

Let's see what historian Kenneth C. Davis tells us about it: "Even though all the players rushed to claim their role in the discovery of America, surely nobody wanted to be recognized for being responsible for starting the traffic of slavery. The unfortunate distinction probably pertains to Portugal, where ten black slaves were brought from Africa almost 50 years before Columbus took his first trip.

But this doesn't mean that the Portuguese monopolized this activity. Soon after, the Spaniards began taking this economical manpower to American soil.

In 1562, English navigator, John Hawkins, started trading with slaves between Guinea and the Western Indies.

By 1600, the Dutch and French were already dedicated to "men trafficking", and by the time the first twenty Africans arrived in Jamestown, aboard a Dutch ship, a million or more black slaves already lived in Spanish and Portuguese colonies in the Caribbean and South America."

It is important to point out that black slaves were not imprisoned in their own land by Europeans, they were actually bought by them from Arab merchants and African kinglets, also black, who sold their enemies and war prisoners—wars which usually started precisely to seek slaves—in exchange for money, weapons, liquor, and other odds and ends.

As expected, black slaves were not only brought from Africa to harvest sugarcane and extract sugar; they were also used for any activity their masters disliked to perform, but had to: Southern United States based its entire cotton industry in slave labor, mining, as well as construction and paving used them regularly, they were even used for household chores, just to name a few. But the truth is that it was the sugarcane and sugar business that contributed to the massive growth of slavery in America.

For centuries, the possibility of extracting sugar from other plants already existed. Egyptians used beet sugar, but only for medicinal purposes. Around 1800, the Germans tried to obtain sucrose from this plant, industrially, but its high cost made them give up on the idea.

It was the French, around 1811, who found profitable ways to produce it, and from then on, both technologies coexisted, countries preferring one over the other depending on their economic interests, agricultural facilities, or political subsidies.

Honey from bees is known and used for thousands of years as a sweetener, as medicine, and as a fermented drink (the Roman's mead).

In the caves of Bicorp, in Valencia, you can see a painting on a rock that represents a man extracting honey from a beehive, while the bees seem to be attacking him.

Hard candies, which were previously made with small balls of honey with fruits, licorice, or other substances, made with refined white sugar, began being manufactured in 1820.

Chiclets, chewing gum with sugar, was first manufactured by William Wrigley, in the city of Chicago in 1892, but in 1893, Wrigley himself introduced the most famous and enduring ones: Juicy Fruit Spearmint gum. Doublemint gum, with the green label, was released in the market in 1914.

It is reported that an American in the U.S. consumes an average of 190 pieces of gum per year.

In 1912, Clarence Crane, from Cleveland, started manufacturing "Life Savers", with the little hole in the middle. A year later, he sold his factory to Edward Noble for $ 2,900 and this magnificent salesman took Life Savers to almost every corner of the world; today, it is a huge company that has gone public.

Nowadays, accused by many as being responsible for our obesity pandemonium, being corn syrup the other guilty party, refined sugar is still used in the composition of many commercial products, and the press often reports about its negative side effects; however, its history, wealth, and immense tragedies, still remains.

Mr. Quetelet

Belgium Adolphe Quetelet (1796-1874) liked the stars and wanted to be remembered for some amazing discovery. That's why he became an astronomer. But in order to study and research astronomy, he needed to measure the celestial space and the enormous astral distance; so what better way to do this than by dedicating a good part of his time to math. And so he took on the task.

While studying math, he realized that not all things could be measured with accuracy; some, like the distance from the Earth to the Sun could, but other couldn't, such as the number of suns in the universe.

This certainty led him to the study of a science he helped create: statistics.

And thinking about statistics, he realized that men, mankind, were very difficult to classify in exact numbers. How long will we live? How tall will my son be? And so on... we had to pursue averages if we wanted to establish some realities.

That's how, in 1870, the "Anthropometrie ou mesure des differentes facultés de L'Homme" was born, a book published in Brussels, dedicated to the study of mankind's measurements.

In this book, countless statistics were presented related to the human body, and Quetelet thought that perhaps the most important thing was to focus on the concept of "the average man".

But no, lost in the pages of this book, he spoke about an index that linked the size of a person in square meters with his weight in kilograms.

He had just seen the light of the famous Quetelet Index, used even today everywhere in the world to define the regular weight or obesity of a person.

No star or asteroid would be named after him, but millions of doctors, nurses, nutritionists, trainers, and other professionals, use his index to classify people, even without knowing who Quetelet was!

OBESITY
CHANGING ITS
SIGNIFICANCE

CHAPTER 16

The Industrial Revolution

The Industrial Revolution, also known by some sociologists as the second wave of human development, being the first the Neolithic Revolution, consisted in a very fast, almost explosive process (in some countries) of technological evolution, and simultaneous economic growth, based on agriculture and craftsmanship.

This sudden evolution unfolded an economy of production of goods, whichever they may be, through "industrial" means, in other words, by means of machineries, large factories, and swarms of workers.

That didn't mean the disappearance of agriculture, farming, and craftsmanship, but industrialization predominated and became the country's main source of income. It even facilitated the industrialization of its own agriculture and of a variety of craftsmanship.

The Industrial Revolution wasn't a similar process across the world. There were pioneer countries, such as England, Belgium, and Germany; then there were countries that were a little slower to reach this point, subsequently catching up and taking a lead, such as the United States, and sadly, other countries have never even gotten there...

As a side effect of the Industrial Revolution, many cities grew rapidly, visible decreasing rural population; the supremacy of royalty, nobility and other idle groups declined, almost disappearing; the "proletariat culture" emerged, which covered an elite of intellectuals, not precisely proletarian, Europeans, whose renowned figures were fat Karl Marx and his follower and advisor Vladimir Lenin. "Industrial magnates" also emerged, self-made businessmen, such as Morgan, Vanderbilt, Rockefeller, Benz, Ford, and many others who raised enormous industrial conglomerates which, gradually and in time, would pass into the hands of the shareholders.

Globally, nutrition improved when agriculture and farming was industrialized, which favored the increase in the production of meats and cereals, and their prices. But in real life, nutritional unbalance became huge.

A tycoon could eat at home, with his family and friends, gourmet delicatessens of incredible sophistication, while workers in any factory around the world would eat the same boring menu, day in and day out, his entire life.

Going through newspapers of the time, specially the English ones, you can see the descriptions and caricatures of obese and arrogant tycoons, wearing a coat and hat, next to famished workers, wearing chaffed clothing and protruding bones.

All of this actually happened and it was repulsive, but the final result was a positive one, and society, as a whole, evolved towards progress, obviously to what we're living today.

It was typical for the more powerful, which were less in number, to continue showing off their bellies, while the poor people, which were more, showed off their thinness.

We would have to wait a century and a half for this scheme to turn around, but the machine was already running...

A good-natured chubby man. Daniel Lambert

There's a lot of discussion about the genetics and other physiological factors of obesity, and here we have a case, extensively published in its time, that seems to prove that not all people are overweight because of excess food and lack of exercise.

On March 1770, Daniel Lambert was born in a rural area of England. He spent his childhood and adolescence hunting, fishing, swimming in rivers and riding horses.

At the age of 21, he substituted his father as the ward in a local prison, but at 23, to his surprise and of his family, he was already weighing 200 kilos, which created great difficulties for him, even to walk.

The funny thing is that the prisoners didn't want him to quit, since his good humor and helpful nature made him an endearing ward. Daniel alleged he ate little and didn't drink alcohol at all.

At the age of 36, he was already weighing 317 kilos and continued growing. Since his obesity prevented him from working, he made the decision to move to London and earn his living working as a fair phenomenon.

As the story goes, the lines to see him were very long, despite the fact that he charged a shilling for admission, which was quite a bit in those days.

His character and kindness never changed. He chatted with everyone who went to see him and everyone respected him. The London press, so sharp and satirical, always treated him with the utmost respect.

On June 21st, 1809, Daniel passed away in his sleep. The wall on his bedroom had to be torn down and it took twenty men to carry his body.

The coffin was reinforced with steel sheets and rods.

CHAPTER 17

Obesogens awaken

For the primitive man, that man without any written history, body fat, being overweight, or obesity was a blessing. It was reserved energy, essential for survival and the enormous physical activity carried out, reserves for maternity and the breeding of neonates, indispensable, first for the herd, and then the tribe.

The Neolithic Revolution, which brought about emerging agriculture, domestication of animals, and the increase of food reserves, degraded the value of obesity as a biological need, but increased it, paradoxically, as a factor of power and social supremacy.

This, without a doubt, was the first change in the significance of fat in thousands and thousands of hominid life on this Earth.

Before the Neolithic period, men lived more than a million years (some scientists assure us that even longer) expanding throughout the African planes and penetrating later Eurasia, America, and the Pacific.

Between the Neolithic Revolution and the Industrial Revolution, some 7,000 years elapsed, but from the latter to the Digital Revolution, which we are living today and is changing us forever, only two and a half centuries have gone by, and in this very short historic period, obesity, like so many other things, changed its significance again, but for the worse.

There are thousands of reasons for this vertiginous change of significance of obesity and fat, directly or indirectly.

Satirizing Churchill, we could say that never have so many things been invented or discovered, by so few, in such a short time.

We cannot, nor do we know, how to mention all of these reasons, but we can pinpoint and comment on some, denominated obesogens, which have played a powerful role in the decay of what the World Health Organization has named Globesity, meaning, an epidemic or pandemic, of global or planetary obesity.

And what is an obesogen? Well, it is a substance, a nutrient, an objective, a habit, a social imposition, a trend, a policy, a technology, any advancement that promotes obesity in human beings. The sum of obesogens that populate an environment, and that infer, one way or another, over the inhabitants of that place, is called "obesogenic environment".

But what is really interesting, is that practically none of the obesogens mentioned here were invented or implanted in our culture and our way of living together, with the purpose of making someone fat.

Moreover, growing obesity has been a surprise for a world that believed, and still believes, that it is making the best to improve the health and wellbeing of a community, and in fact, it has achieved that in other aspects. Only in the last forty or fifty years, a sector of the population has become aware of the phenomena, and now, just a few years ago, campaigns to confront this problem have been seen.

Let me tell you a brief story.

It was 1968 and the world was fighting over conflicts and apocalyptic predictions. The Cold War between the two superpowers, the Soviet Union and the United States, with its indications of a final nuclear confrontation; the conflict of Vietnam devouring more and more resources and lives; the beginning of the student protests against the Vietnam War; the May revolts in France, which put the government of General DeGaulle at the border of collapse; the Prague of Spring and its tragic finale under Soviet tanks; the Arab-Israeli military tension a year before the Six Days War; the growth of the hippie culture and drugs; the Latin American guerrilla sponsored by Cuba and its colophon in the death of the partisan Che Guevara in Bolivia; the hot summers of the North American Civil Rights Movement, and.

Finally, a book from biologist Paul Ehrlich entitled "The Population Bomb" sees the light. This book speculates, based on governmental documents from FAO and the United Nations, about the certainty, if extreme measures were not taken, that humanity would follow the path of massive famine that would kill hundreds of millions of people in the most economically backward countries, in turn generating more unstable politics, mass protests, and eventually new wars.

Ehrlich based his book in the fact that agriculture production of foods grew arithmetically, while human population grew geometrically (a return, more moderns and informed, to Malthus).

From the mathematical point of view, Ehrlich's argument seems impeccable, and that's how other researchers recognized him, but...

While this scenery of calamities occupied the front-page of all newspapers and television headline news, a man of few words, also with a solid biology and plant genetic training, traveled to India and Pakistan from Mexico, where he had worked, taught, and learned agriculture since 1943, to advise these countries on agriculture techniques that would help them confront the problem and also change the world.

His name was Norman Borlaug. Using genetic engineering techniques, chemical fertilization, intensive irrigation, and the elimination of undergrowth, he was able to get wheat sown land, which in 1950 produced 750 kilograms of cereal per hectare, to produce, by 1970, between 3200 and 3300 kilograms per hectare. This was called the "Green Revolution" and won Borlaug a Nobel Peace Prize. Ehrlich predictions were forgotten, at least for now.

The images of African children consumed by hunger, anorexic, wasted, with protruding bellies (because of parasites) and about to die, were gradually fading after the Green Revolution. Without a doubt, Borlaug and his followers saved the planet, but when the consumption of cereals and their derivatives, which are hypercaloric nutritional elements, increased, they were also unintentionally aiding in changing the significance of malnutrition.

Malnutrition without anorexia, even being overweight or obese, but malnutrition after all, since the contribution in vitamins, minerals, and amino acids, essential to diets based on cereals, is not sufficient if other sources of nutrition are not consumed.

A place to eat. Restaurants

Travel lodges, where people could eat from a common pot, drink the house wine, and then feed the horses and spend the night, were already known, at least, during the Middle Ages. Monasteries and convents also offered hospitality to those who had no choice other than to travel from one place to another during times when they couldn't even dream about tourism or pleasure trips. The customary thing to do, until about the beginning of the 18th Century, was for travelers to carry their food and wine in their satchels.

The first restaurant, with the concept we understand today as such, was founded by Parisian Antoine Boulanger in 1765. His basic dish was soup, but some days of the week, visitors could choose some kind of roast.

In 1782, Antoine Beauvilliers, who subsequently would become famous because of his recipe book, founded "Le Grand Tavern of London" in Paris, where he established standards that are quite common for us today: time for opening and closing the premise, personalized customer service with your own servant or the owner himself, a menu with daily entrees and certain specials (for special clients), some privacy, and the check at the end.

English philosopher and writer, Samuel Johnson, absolutely opposed the concept of eating in individual tables, very modern by the 18th Century, alleging, and with his own words: "that this very ultramodern atrocity will end with civilized coexistence." Interesting observation, right?

In 1800, there were already dozens of restaurants in London and Paris. Restaurants were a place for gathering and conversation, of course, also a place where people with financial means could get food.

But soon, shrewd chefs realized that serving food would be a good source of income if they increased their clientele.

In the Old West, Chuck Wagons emerged, taking food and drinks to farmers—cowboys—who gathered the herds in the open fields. A saloon, so typical in western films, had to do much more with prostitution than nutrition.

The first meals served in the air occurred in German airships, but the zeppelin catastrophe of Hindenburg cut this custom short.

Nurse Ellen Church was the first flight attendant in a commercial flight, on May 15, 1930. She offered coffee and sandwich to passengers, but the idea died out until 1935, when American Airways introduced serving hot meals to passengers, a practice that still exists, with its ups and downs, today.

And what about today? Ethnic restaurants, pubs, carberies, bistros, bouchons, oyster bars, meaderies, juke joints, with their fascinating history of the U.S. deep south, the greasy spoons, themed restaurants, underground, Cuban palates, those constructed in ice, the nudists, Puerto Rican "lechoneras", raw bars, hostels, floaters, and so on; you can fill pages and pages of them.

How about we go to a nice restaurant now.

Will you join me?

CHAPTER 18

Ingested obesogens

If obesity is made up of deposits of fat in body cells called adipocytes, which can grow many times bigger than its original size, it is reasonable to think that excess consumption of animal and vegetable fat are the cause of obesity.

It actually doesn't happen like this exactly.

Fat deposited in adipocytes don't come from the outside; the body produces it, especially in the liver, as reserve, fundamentally using carbohydrates or sugars (it also uses proteins and fats, but with slower metabolic speed).

This explains the importance of excess of calories and the excess of carbohydrate in your diet as a primary source of obesity. However, we must clarify that other factors also play different roles in this complex mechanism.

We've already discussed sugar, rice, chocolate, bread, and other nutrients (they are nutrients, even though they may be harmful in excess), directly related to obesity in today's world.

Since this book is not a physiopathology treatment or an exhaustive gastronomic assessment, we will just quickly mention some agents, such as industrial candies (mass-produced baked goods), TV dinners, alcohol: some good in moderation, such as red wine or an ounce of whiskey, and others not so much; sauces, pastas, chips, and so many other gastronomic products that have turned into the source of much of humanity's nutrition (filling up the tank, actually), especially the younger generations.

Let's briefly take a look at the history of some of the better known products, and most attacked, even though this attack, in a typical love-hate relationship, does not prevent their increased consumption.

Corn and its derivatives

While this little book is being written, in the central plains of the United States, more than 300 million tons of corn is being produced with very sophisticated technological methods. No other country has been able to achieve this amount, but this wasn't always so.

Corn, like potato and cacao (chocolate) is from pre-Columbian America. There is evidence that about 7,000 years ago, corn was harvested in Mesoamerica, specifically in the region where Guatemala seats today. The Mayans knew corn very well and abundantly used it, to the point that these civilizations became known as the corn culture.

It's funny that because of the agricultural manipulation of grains, corn lost the ability to self-reproduce, a rare genetic phenomenon.

The fact that nowadays we have no wild corn, turning the plant into an almost perfect cultural "case" is quite interesting.

To make corn even a more relevant plant, its active genome—some 55,000 genes—is almost double of that of humans.

Colonizers would soon take corn to Europe, and soon after it became indispensable for humans, and even more for animal nutrition.

Pilgrims and settlers of the 13 U.S. Colonies grew corn, but they also received it from Europe, which proves its rapid expansion and penetration in European culture.

The "star" crop of the Industrial Revolution was the corn, and this was due to its high production, which turned it into "food for the indigent" by excellence, and it also continued being an intermediary producer of meat for human consumption, especially poultry and pork.

By 1947, almost all filmmakers in the United States received an extra earning for the sale of popcorn. Today, most of the profit earned by movie theaters comes from the sale of popcorn and other snacks.

But let's travel a bit back in time. John H. Kellogg (1852-1943) was the loyal follower of Mrs. Ellen G. Harmon (1827-1915), one of the founders of the Battle Creek Seventh Day Adventists church in the Midwest.

This sanctified lady avowed that when she was young, she had a vision, where God told her that breakfast was sacred and that corn, which had fed the pilgrims, constituted her main contribution.

Kellogg, who blindly believed the sermons this saint told him, got her to sponsor his idea of processing corn to make it more appetizing and manageable.

And so, Mrs. Harmon and her entire community did it, kicking off the industrial processing of the cornflakes, one of the most profitable and productive U.S. businesses in history.

In the 1970's, the Japanese invented corn syrup, which together with refined sugar, are the worst of the worst for the world's obesity epidemic.

Today, we can't envision the food industry without corn syrup, one of the most versatile obesogenics (and for some very harmful in other aspects) of industrial gastronomy, a synthetic product that has been substituting sugar in the production of sodas, many sauces, and even numerous solid foods, where we can't even imagine this type of ingredient being a part of.

Sodas

During the 19th Century, Southern U.S., due to the heat, agricultural work, and low income residents, undoubtedly lead the production and consumption of soda water.

It seems then very natural that in May of 1886, an Atlanta pharmacist, John Syth Pemberton, banged his head against the wall to find a soda that tasted better than the others: the perfect tonic water and the ideal stimulant.

Pemberton, who was pretty much up-to-date in pharmacology, knew about coca leaf. The French Wine

Coca was his first intent, and a copy of the "Vin Mariani", a wine manufactured by an Italian that was made with coca and sold really well, but was banned, not because of the coca, but the alcohol.

But Pemberton, an innate fighter, didn't give up. He mixed distilled coca leaves, macerated in kola, sugar, and soda water. Now, he just needed a name.

And the name came from the mouth of his friend and accountant, Frank Robinson, who also lent him money several times. Naming it Coca and Kola seemed too obvious, but Pemberton thought that Kola with a "C" would look better: Coca y Cola; why not Coca Cola?

When someone has a good idea, people usually say that person invented Coca Cola, in this case, Pemberton (and Robinson) really invented Coca Cola.

But unfortunate Pemberton died of stomach cancer a year later, not without first selling the Coca Cola patent—he had legal problems with Robinson—to Asa Griggs Candler, another businessman from Atlanta, for $283.29.

Now this man really invented Coca Cola!

In exactly 31 years, Candler took Coca Cola out of pharmacies and turned it into a national beverage. He sold it to a group of bankers in 1919 for 25 million dollars.

It was the biggest negotiation ever made in Southern U.S. since before the Civil War. And the rest is history. Coke, as indicated in the stock market, is a constant presence in the entire world, even in space!

Pepsi Cola was born around 1890, as the "Brad Beverage", because of its creator, the pharmacist from North Carolina, Caleb D. Bradham. In a few years, attacking the hypochondriac side of so many people, he changed its name alleging it was really good for the dyspepsia: Pepsi... Cola.

In 1950, Alfred N. Steele, who, by the way, was married to Joan Crawford, saved Pepsi Cola from bankruptcy and made it big, almost alone, reaching the level of sales and fame of Coca Cola.

In time, just like Coke, Pepsico, which is how the company is publicly known, developed an inventory of products that go from bottled water, one of the most incredible businesses of the last years, to other sodas, energy drinks (high doses of caffeine and corn syrup in a can), tea, fruit juices, etc.

Dr. Pepper, who has no cola, was born in 1885. 7-UP was created in the midst of the 1929 crash by Charles Leiper Grigg (7 because of its seven components). And so the history continues, almost infinite, of sugared drinks.

Today, thousands of them are manufactured and sold; each country has its own variety of sodas, but none, from Asia to the Patagonia, has been able to surpass Pemberton's absolute success.

Oh! By the way, one of the biggest problems in Mount Everest, is the rising number of Coca Cola cans and other sodas lying on its foothills, an ungrateful memory of so many successful and unsuccessful climbs.

Hamburgers and others alike

Between 1200 and 1300 CE, fierce galloping Mongol riders, who ravaged the plains of Asia and boldly penetrated Eastern Europe kingdoms, nurtured themselves without even getting off their horses, with raw meat, cut into strips, which the Europeans called "steak tartare".

This steak became inherent of the regions of the Baltic Sea (Finland, Estonia, Latvia, Lithuania, and Northern Germany), where they adapted it by adding new seasonings, salt, and onion.

In the Baltic port of Hamburg, the largest and most important of Germany, a special and quite popular new taste for beef emerged. But it was grounded and mixed with eggs and chopped onions.

At the end of the 18th Century and beginning of the 19th, many Germans immigrated to the United States in search of work. They began their journey in the Hamburg Port. And this is how this "Hamburg-style beef" came to this thriving American nation.

The new flavor took on two different paths; one was refined, accepted in luxury restaurants, such as the Delmonico's in New York in 1834, and another popular, consumed in large quantities by seamen in the port of the Big Apple and German immigrants from the Ohio River Valley.

But something was missing.

Bread was missing. And the American mythology tells us (and myths always have some truths in them!) that a 15 year-old kid, Charlie Nagreen, who sold food at the

Wisconsin State Fair, put "Hamburg-style" beef between two pieces of bread so that his customers would not get their hands and clothes dirty with the sauce.

Since it was a big hit, he named it "Hamburger".

Other stories are told, but this is the one I liked the most, and besides, it has the perfect name!

However the case, in 1904, hamburgers were already trendy in many spots around the United States, even reaching the Pacific Coast.

In 1916, a hamburger chain of restaurants already existed: White Castle, founded by Walter Anderson. And by 1930, there were some 150 hamburger joints in 10 states.

In 1937, two brothers, Richard and Maurice McDonald, taking advantage of the popular automobile boom, opened a drive-in restaurant in Arcadia, California, where they sold sandwiches, hot dogs, hamburgers, and shakes.

They did really well, but they noticed something strange: they expected hot dogs to be more popular, but no... hamburgers were sold more, with a 90% advantage rate!

So they made a risky move.

They eliminated the drive-in and established what they called a self-service restaurant. Customers would serve their own hamburger, ordered their shake, paid, ate, and tossed any leftovers in the trash. The first week was chaotic, but they persevered, and everything started to improve.

By 1951, they were selling more than a million hamburgers a year.

In 1954, a kitchen equipment salesman, Ray Kroc (1902-1984) received an unusual order from a small restaurant called McDonald's: 8 mixers. His good nose for business led him to deliver the mixers personally, and check out what was going on at that restaurant.

It was love at first sight. Kroc, who was a risky man, asked them if he could open a restaurant with the same name, but far away from them. They accepted.

The rest is history.

In 1955, Kroc opened his first McDonald's in Des Plaines, Illinois, making $366.12 on his first day. By 1959, he had over 200 hamburger joints in the United States, and in 1961 he begins his attack to conquer the world. At the beginning of the 21st Century, Thomas Friedman exposes his "Golden Arch Law":

Friedman states that "no two countries with McDonald's within their borders have ever been in war since having a McDonald's." We'll leave the analysis to your discretion.

James McLamore and David Edgerton founded Burger King, in Florida, in 1954. Harlan Sanders, a 65 year-old man, almost bankrupt, used $105 from his Social Security check to promote franchises of his chicken recipe: and so Kentucky Fried Chicken was born in 1939. Glen Bell opened his first Taco Bell in Downey, California, in 1962. Wendy's was born from the hands of Dave Thomas in Ohio, in 1969. Subway was born in 1965. Quiznos in 1981.

What's next?

Pizzas

Historians are always seeking references for everything. Pizza, one of the most popular dishes in the modern world, has plenty of them!

Whether bread accidently burnt thousands of years ago, or flour was placed on hot stones, whatever. What we know with certainty is much more recent.

Focaccia, a flattened and thin flat bread, covered with aromatic herbs, was quite commonly eaten in the Middle Ages (and still today), especially in the south of the Italian Peninsula. Everything points to the assumption that focaccias actually evolved into pizza.

But before focaccia became pizza, there was an ingredient missing: tomato. Tomato came from America a few years after its discovery. At the beginning, people feared them. They said tomatoes only served as an aphrodisiac or that it could poison a person's blood; so people only used them as ornaments.

Pizza was Italian, but the best pizza was born in Naples, where they were sold on the streets and plazas, made right there, as it should be, by street chefs who used portable burners.

In 1830, Antica Pizzeria Port'Alba opened its doors, the first true pizzeria recorded in history, and which still exists today.

The vast Italian immigration of the 19th Century to America was already on a roll, and in 1905, Mr. Gennaro Lombardi opened the first pizzeria in Little Italy in New York. Then followed Boston, Chicago, and the world. Pizza Hut opened in Kansas, in 1958. Domino's Pizza was born in 1960. Sbarro in 1967.

Papa John's was founded by a student, a pizza delivery boy for another company in 1980. Telepizza, created by a Cuban in Spain, in 1986.

And we could go on....

Junk food and empty calories

Trash food, fast food, or junk food, is nothing more than a gastronomic and sociological concept, but a concept supported by the daily practice and observation of a phenomenon that by dint of being more and more common, became to be "normal" or socially accepted.

If a hamburger or a pizza were made with real, first-class, fresh products, with all the amino acids and proteins of beef, the necessary fats, vitamins from tomatoes, and minerals, such as calcium, from cheese, they wouldn't be considered junk food, despite having high calories or consumed in large quantities.

In any case, we would be discussing the amounts ingested and not their composition.

But the problem is that in order to attend to the immense current demand for these products, their ingredients have been degraded, distorted, frozen, mixed, stored, contaminated, to the point of being considered by some gastronomes as UFOs (Unidentified Food Objects).

It seems really hard to figure out the exact date and origin of the ingredients of many of these commercial products that drown our cities.

If we add to this the fact that among their components we can find highly saturated fats (trans fat have been banned from some countries), enormous amounts of salt, coloring and thickening agents, the damage in the organism caused by their repeated consumption (arterial hypertension, vascular obstructions, diabetes, some neoplasms) goes beyond obesity.

Junk food, as a concept, is linked to serious social and cultural problems in current urban life: increase of the so-called "vagabond food", loss of family commensality, abuse of social commensality, family dysfunction, lack of free time, loss (extreme monotony) of taste, above all in children and adolescents, addiction, lack of food quality, and a very long etcetera...

If we take a look at international statistics about weight, obesity, and type II diabetes mellitus, the relationship between extension and growth of world consumption of this type of food stands out, and the exponential jump, in the last 30 or 40 years, of these pathophysiological conditions.

And what are empty calories?

Well, it's nothing more than the definition of an edible or drinkable product that contributes calories to the human body without contributing any type of nutrients.

Pure water has no nutrients (in reality, it has some necessary minerals, but very little), but it also doesn't contribute calories, and that makes it healthy. Sodas, on the other hand, contribute large amounts of calories, from sugar and corn syrup, and no real nutrient.

If we add to this the large quantities of sodas consumed by our population, especially by children, we will once again see the close relationship between obesity and consumption

Gastronomy and Gastro-anomy. A marginal note

In 1940, coming up with daily or weekly menus was very easy. The selection of each type of food was reduced to three or four options, and in some cases, not even that. A tub of butter was nothing more than a tub of butter, and that was the one bought consistently, and consumed at home with delight.

Nowadays things are a bit different. At any supermarket, a customer will find 30 or 40 different sizes, presentations, and brands for almost every product.

A tub of butter may now be organic, red, colorless, fat free, salty, unsalted, kosher, natural, goat or other mammal, soy, vegetarian, gourmet, expensive, cheap, and so forth.

And the same thing goes for almost every other product consumed by inhabitants of the most developed countries and many of the ones that aspire to be developed.

In the past, people guided themselves by their available financial resource, family customs, traditions, and habits of their human and social group. Food choice was usually done prior to shopping, without the need of a particular intellectual or decision burden. But all of this has changed with dietary transition, extreme competing, and industrial commercialization of foods.

Social evolution has changed the cultural and social patterns of an individual. Family, work, and social environment, traditions, information, and many other aspects have completely distorted old traditions and customs, but above all, the field of possibilities of election has opened an immense spectrum.

Individuals must now choose, they are required to choose, whether they like it or not, and this constant choice is what sociologist Durkhem denominated "anomy".

Anomy is nothing more than a state of resignation, of weakness, volition at the face of such compulsion that is too constant and permanent.

Individuals begin to delegate their selection to trend, media propaganda, opinions of others, commercial bombardment, "discounts", labeling, advice from nutritionists and specialists, whatever is said to them. Anyhow, they confer their judgment to the changing and voluble opinion of the masses.

It's really incredible the stress this "freedom" in food selection can generate a person.

Never before has mankind had such freedom to choose and at the same time more tied to decisions and opinions of others to their own will.

It is this state, continually growing and more intricate, in current social order, that has been denominated, as a comparison between traditional gastronomy and this decisional chaos, gastro-anomy.

CHAPTER 19

Obesogens that are not ingested

It is endlessly said that there are two reasons for obesity:

Caloric hypernutrition—ingested obesogens—which we have already mentioned to some extent, and a sedentary life, absence of relatively vigorous muscular movement that generate some caloric expenditure, capable of balancing caloric intake.

A sedentary life is almost never and isolated event; it happens in an environment that "prevents" a person from moving or minimizing movement, and this environment is made up by obesogens that are not ingested, or in other words: an obesogenic environment.

As technology develops and grows, and urban life becomes the norm, while rural life is developed, the obesogenic environment starts limiting people, especially in the most developed countries, in their physical activity, at the same time as it promotes, directly and indirectly, high-calorie nutrition.

From a strictly scientific point of view, several factors play in obesity—genetic, physiological, endocrinologist, psychological, and neurological—that go beyond high-calorie nutrition and sedentary life, but it is these last two, becoming in indissoluble part of modern life (we should actually call it postmodernism), the ones that have triggered the worldwide pandemic development of obesity: the aforementioned globesity.

There are thousands, perhaps hundreds of thousands of non-ingestible obesogens.

Any object, product, or gadget that makes life easier, and therefore, leads us to decreased muscular activity, is a potential obesogen.

An elevator promotes the uselessness of stairs; a crane minimizes the physical task of lifting weight; the metro eliminates transportation by horse; pipes and engines make it unnecessary to carry water in a bucket, and so on.

Of course it is quite clear that this is not really bad if we become aware that we need physical activity, exercising, battling a sedentary life; but the reality is that comfort is relaxing, and what in the past was done habitually, now it requires more effort, the effort of determination, which sometimes, because of society or socialization is almost impossible.

Let's briefly review, and only as an anecdotal reference, some of the most common and well-known obesogens that are not ingested.

Automobiles

Wagons, chariots, carriages, cars, sedans, wagons, gillnet, and all kinds of carts pulled by horses or other animal or slave force, were used from the beginning times.

In the 18th Century, two artifacts claimed to be the first "automobiles": the Gugnot tricycle (1771) with a steam broiler in front that prevented visibility; and the Murdock tricycle (1774), with its smokestack in the back. Both endured as simple curiosities.

The idea of constructing a self-propelled vehicle continued during all of the 19th Century. Electric, gas, and oil engines were tested, and even Dunlop devised his first tires with rubber bands, but the first truly efficient gadget was created by Butler in 1884. The car died almost as fast as it was born because the English authorities opposed the fact that it moved more than 4 miles per hour.

At the end, it was the Germans who had the upper hand. Gottlieb Daimler perfected the cylinder and the piston of the steam boiler and constructed the first motorcycle. Karl Benz worked with gasoline (a byproduct of oil, which had no practical use at the time) as fuel, which allowed for the construction of the first automobile between 1885 and 1886. Immediately after, as it happens when a door to the unknown is opened, emerged those cooled by water (radiator), the clutch, the differential, dual front wheel, the roof, etc.

Back then, the Americans had little to say about this topic... until a man named Henry Ford (1863-1947), an engineer of the Thomas Edison electric company, took on the task of constructing an automobile, which he did in 1896.

Ford started copying the French and Germans, but like a good American, he visualized aspects that nobody had taken into account before. Until that time, automobiles were toys for the wealthy, but they were manufactured one by one, and none of them equally.

Ford realized that the potential business was in that each family should own a car, and this required a change of mentality and the development of a technology to lower production cost.

As a curious note, it was Edison who convinced him to use fuel engine instead of electric power. I wonder what would have happened if Ford hadn't listened to him?

In 1903, he founded "The Ford Motor Company" with $28,000, of which only 25% were his. In 1908, he released in the market the T Model. In 1910, he placed the engine in the front of the car and in 1913, he patented the interchangeable or spare part system and the assembly line, which was his biggest and most renowned contribution to industry, and not only for automobiles!

By 1926, he had sold, at reasonable a price, 15 million Fords.

After creating, through franchises, the mass sale of parts and fuel, Ford modified the image and way of life in the United States. Henry Ford didn't invent the automobile, but he did so much more than that; he created, with his industrial and commercial vision, a world that cannot possibly reverse.

What subsequently occurs is pure history...

The Office

An Egyptian scribe, devious and manipulative, as described in almost every Hollywood film, was an office clerk. Can any magnate and statesman today boast a secretary as Alexander of Macedon: none other than Mr. Aristotle. Rome created the word *"officium"*, and it also invented the bureaucracy, formed, in many occasions, by Greek slaves.

The Middle Ages brought us copyists, men with great patience and unparallel perseverance, to whom we owe, among other things, the knowledge of the great Greek poets and philosophers. But it was the colonial system and Industrial Revolution that allowed for the development of the "office", more or less as we know it today.

The contribution of the modern office to sedentary life, and conditioned social power, is much bigger than you can ever imagine at first glance. Minuscule cubicles, swivel chairs, intercoms, computers, photocopiers, all of them conspire to making a person inactive, and being, distractedly, consuming calories the entire day!

Ah! And the person arrives and leaves the office using the elevator!

Television and its environment

Some sly journalists have compared television to God. If you think about it, the television is also everywhere, but it doesn't have as much power as God.

The history of this invention is long and complicated, partly because it can't be attributed to a single person or institution. From Michael Faraday and Joseph Henry in the first third of the 19th Century, going through to Caselli, Smith, and May, Gerogeo Garey, Goldstein (the one from the cathode rays), Edison, Tesla and Graham Bell, up to Paul Nipkow, Ferdinand Brand, the Russian Constantin Perskyi (the first to use the word "television"), John Logie Baird and Philo T. Farnsworth, the ingenious Californian who made it operational, the television is the product of the minds and intelligence of hundreds of people.

On September 7th, 1927, the president of the United States, Herbert Hoover, officially inaugurated it. In 1953, NBC aired color TV, and in 1975 satellite broadcasting (SATCOM) was introduced. Quite a fast advancement for a technology system that has completely dominated the planet.

In 1955, the company Zenith introduced the remote control (it was still a flawed piece of equipment and not very practical) and the road to absolute immobility begins. A reclining chair, TV dinners, flat and huge screens, surround sound, cable and satellite channels, cooking channel, specialized TV chefs...

In conclusion, the obesogen television environment.

Refrigerators and kitchens

For thousands of years, cooking was always repetitive and boring. Probably thousands of years elapsed from roasting a piece of beef right on the fire to the intelligent action of introducing it in a pot of water and heating it.

There is no definitive archeological evidence that houses from the very first civilizations had a kitchen, except for large palaces. People cooked in plazas or open places in the villages. In Greece, people would cook outdoors, and while they cooked, they would chat and discuss with other people.

In Rome, citizens with good financial status already had kitchens, but the underprivileged had community kitchens.

The Middle Ages brought two very important advancements: the chimney, which allowed smoke to escape, and iron pots. DaVinci, who practically did everything, also worked in a system to propel smoke upwards, eliminating itchy eyes and throat that persecuted mankind for thousands of years.

The blender, or mixer, was patented in 1922 by Stephen Poplawski, but it was Fred Waring who made it more popular, and became rich with it.

The microwave was discovered, coincidentally, by Engineer Percy Spencer, while he tried to improve the primitive radar of World War II.

The Chinese, who always surprise us, developed a way to preserve ice, using storages lined with wood and double boxes, some 3,000 years ago.

The first refrigerator was invented by German engineer Carl Von Linde, in 1876, parting from a technique he had designed some years before for the manufacturing of Guinness beer.

In 1916, Alfred Mellowes manufactured a metallic cooler, hermetically closed, which ended up being, essentially, our current refrigerator. However, after borrowing money and confronting several problems, he sold his patent to

General Motors, yes, the same as the cars, which created the company Frigidaire.

Later, General Electric gets into the business, and as always... the rest is history.

So here you have, men unintentionally created techniques and mechanisms that made our lives more pleasant and enjoyable, but they also kept us away from fresh and natural foods.

Supermarkets

It was U.S. businessman Clarence Sunders, while researching on a way to increase the productivity of his employees in a small retail food store, who came up with the idea of packaging food and putting it at the reach of his clients, in a larger space.

Over time, he acknowledged that he had done this to reduce the number of employees he had (a good example of a commendable achievement done with a malicious intent). On September 16th, 1916, he inaugurated the first supermarket, Piggly Wiggly, in Memphis, Tennessee. He put that strange name with the idea, quite ingenuous, that people would go in just to check out what was inside.

A month after the inauguration, he patented the word "supermarket".

There were other successful attempts, but it was after World War II when an explosion in the growth of supermarkets occurred. Large companies began to invest in them and they expanded to Mexico, Canada, and Europe.

In the 1950's, Sam Walton brings back the concept of a grocery store (that sells a little of everything) and expands it worldwide, never imagined before. Wal-Mart is the largest retail store in the world.

In 1938, an associate of Piggly Wiggly, Sylvan Nathan Goldman, realized that customers were using their arms and hands to hold their merchandise, and once they couldn't hold anymore, they stopped shopping, something that now seems so obvious but it wasn't back then.

He got together with a mechanic, Fred Young, and they manufactured a frame with roller skate wheels, over which they put a folding chair and two baskets. A supermarket cart was born. In 1947, another mechanic, Orla Watson, invented the telescopic cart (one enters inside the other), configuring the typical supermarket cart just like we know it today.

Digital world

If this book had been written thirty years ago, say around 1980, the mention of computers and computing systems would have been slightly less than formal.

Broadband internet, the internet, world wide web, personal computers and laptops, tablets, e-readers, video games, social networking, online encyclopedias, search engines, digital agendas, face to face communication via satellite, smart phones, geopositional satellites, and so many other factors that made our world another world, we would have simply not mentioned them because they didn't exist.

Making history out of this, with agreements and disagreements, with their spectacular leaps and grand failures, with figures like Bill Gates, Tim Berners Lee, Steve Jobs, Steve Wozniak, Paul Allen, Mark Zuckerberg, and dozens of others, would require them, not just this little book, but several large volumes that already exist and quickly get old.

Never before had a generation, or two, developed a technology and a form of communication with such penetration throughout all levels of society that could change the costumes and even the way of living of practically an entire world, even those who refuse to adapt to the digital environment.

Therefore, let us record what digital revolution has meant for humanity and, as a side effect, for obesity.

Let's say that these few paragraphs are the inverted expression of a phenomenon that has surpassed all others, and that in the terrain of overweight and obesity it could only compare to the effects of the automobile, but covering more space and more time.

If we already said that a non-ingestible obesogen is one that, creating comfort, reduces the need for physical activity, then the complex digital world is the king of the obesogens, to a level never dreamed of before. However, they came to stay and to develop to unimaginable heights, therefore, let's learn to live with these things and seek formulas, even within the phenomenon itself, to help us through this obesity crisis.

The digital revolution already changed the world, and will continue changing it, and will surely change human beings.

Are you doubting me? Wait seven or eight years.

A "fat" prize

Lucian Freud was born in Berlin, in 1922, but he was naturalized British. His father, Ernst, son of the very famous psychiatrist and neurologist Sigmund Freud, creator of psychoanalysis, decided in 1933, when Lucian was just 11 years old, to take him out of Germany.

He figured that the arrival of Hitler in power, would not offer a good future for the Jews in Germany, and that's exactly what happened.

Being a painter was Lucian's aspiration, and from a very young age, he started studying and training with great discipline and concentration, but his artistic form was a bit archaic.

He enrolled in several schools, but his true creative source was the streets and some painter friends from the School of London: Mason, Michel Andrews, Bacon (the one he adapted the most to), Kitaj, and some others.

He chose to be a portrait artist, in a time when portraits were not trendy. But he wanted to paint real people, not famous personalities or commissioned jobs.

Another of Lucian's characteristics was that he painted his model, persons, or animals, with their name and in a realistic fashion. He once said: "I paint people, not for what they want look like, but for what they are."

In 2001, he broke the rule of not painting renowned people (he had done this once in a while), and he painted King Elizabeth II of England.

The painting was a scandal, since the queen is portrait exactly as Lucian saw her: very old, messy, and insignificant.

Lucian sells his paintings now for millions of dollars, but his record was reached in 2008, when the Christie's House auctioned his painting "Benefits Supervisor Sleeping" for 33.6 million dollars, the highest amount sold by a live painter.

And who did he paint on that canvas? His friend, Sue Tilley, a relatively young, 400 pound, tax subsidy supervisor.

AH! And she didn't make a dime posing... while she slept in his sofa!

CHAPTER 20

The American Paradox

The transition from a nutrition with natural and fresh factors, although not always necessarily healthy, typical of a pre-industrial society, towards a nutrition based on prepackaged, precooked, frozen, and processed products, or even of genetic design, with preservatives, antibiotics and other substances, that has been growing every day after the Industrial Revolution, until reaching the most absolute, at least in the Western world, throughout a period of 40 years, has been called "nutritional transition".

The United States has undoubtedly been the leader in nutritional transition.

This does not mean that this transition has been the result of a design created by Americans.

No, it has simply been a product of the inventive and business competence of many people working in different spheres, some of them so distant that nobody could have prevented the final result.

Almost all processed foods or designs have been invented by Americans, and the same thing occurs with the non-ingestible obesogens.

Of course, this can explain the alarming increase in obesity in the United States, a phenomenon that has been unstoppably increasing since the 1970's.

The interesting thing is that the United States has also led advances in health and medicine throughout the 20th century.

American universities are home to a good number of Nobel Prizes in physiology and medicine, both born in the country as well as foreigners, and the U.S. government and healthcare companies spend approximately 16% of U.S. GDP in health and medical treatments, figures which no other nation can reach.

Americans also are at the forefront of diets, organic foods, low calorie nutrients, "natural" recipes, sale of concentrated vitamins and minerals, associations and gyms, home exercise equipment, and the huge "fitness" business.

Summarizing:

The United States ranks number one in the world in the production of processed foods and food consumption per capita, which makes them the most obese country on the planet, but they also spend more on healthcare and fitness than any other country.

This incongruity, which possibly could be explained by the very American lifestyle, but that definitely needs a more elaborate and scientific understanding, has been named the American paradox.

Neophobia and Neophilia

I know a person who loves to try new things, especially anything related to food. This person travels a lot, and can eat anything: tripe in Argentina, water buffalo meat in Africa, Korean dishes in Brazil and a fairly edible *pabellon criollo* in a beach in Thailand. This person is an example of what psychologists have named a "neophiliac", someone who seeks new flavors, aromas, colors, and textures at all times.

The author of this humble book is completely the opposite: regarding gastronomy, I rather eat what I know, what I like, and what I've loved for decades. In the hypothetical case of being the sole survivor in a deserted island, I could survive my entire life eating white rice, sausages, and tomatoes, or at least until someone rescues me.

Anyway, besides being a disaster, I am also what specialists refer to as "neophobic", in other words, a person who prefers not to take risks trying new things, or the unknown, when it comes to culinary tastes and preparations.

Both characteristics are very specific of living creatures, and most surely, as indicated by a variety of investigations carried out in the last 20 years by prestigious scientific institutes throughout the world, they are imprinted in the genetic code—genome—of different species.

So why is it then, that in certain aspects, one predominates more over the other?

Well, that's because both are necessary for maintaining life, and in fact, both are manifested in all live creatures at one time or another during their development and evolution.

A panda bear is completely neophobic when it comes to food, only eating tender bamboo, but very neophilic when it comes to humans. Sunflowers are highly neophilics to the sun and very neophobics to the lack of water. Christopher Colombus was a neophiliac, obsessed with species, gold, and Moluccas, but somehow neophobic in his personal relations, even though he had to be surrounded by people and appeal to them to achieve his goals.

Even though phobias and philias can be applied to almost anything, what we're interested in here is obesity, and in this case, children's obesity, and neophobia or rejection to the incorporation of new foods plays an important, however very unnoticed, role in the obesity pandemic that a big portion of this planet's inhabitants suffer from.

When man was nothing more than a biped humanoid, the food he could find, collecting seeds, pulling fruits, or scavenging any remains left by other wild animals, bigger than him, were very scarce and many times highly dangerous.

An unknown food may be a relatively tasty nutrient, such as an almond or a piece of fresh meat, but it could also be a lethal poison, such as wild bitter cassava or rotten and contaminated meat from any animal.

And here's where neophobia plays an important role. If through trial and error you would try contaminated food, and you would survive the experience, the taste and smell of that food would never be forgotten.

In time, and through generations, taste for the unknown was setting in, thus strengthening the barrier of defense when faced with the unknown and potentially dangerous, in other words, nutrition neophobia was establishing itself. Neophilia is also important, and thanks to it we cook today our food and enjoy French haut-cuisine, but that's another story.

How about current obesity? Well, most young children present some degree of neophobia after the age of two, which translates into the rejection of incorporating new foods into their diet, especially seafood, vegetables, salads, and fruits, a phenomenon that has already been noted by U.S. psychologist, William James, in the 19th Century, and which today pediatricians name children anorexia.

But nutrition neophobia has, almost always, an exception, and that is the sweet, or relatively sweet, flavor.

If parents, with the help of a pediatrician, would understand this stage in the development of their children, and adequately manage it, neophobia, in good measure, would stay behind, nutrition would be more adequate, and the possibility of being obese in adulthood would decrease.

But if you take the easy road, and feed your children the same-old thing that they already know and accept: pasta, pizza, rice, mash potatoes, sweet juices, ketchup, sweeten milk, crackers, sodas, all types of snacks and later on, fast food, ideal for parents who are in a rush, we can clearly start seeing the relationship between neophobia—idly managed—and obesity.

CHAPTER 21

The decline of ethnic food

Chatting one day with a friend, he was telling me about his preference for a typical Thai dish, a country he knew very well, and he ended up telling me that the best Thai food he had ever eaten was in a restaurant in Caracas: "That one was really good!"

The average American adds ketchup, ground pepper, and spicy sauces to most of their dishes, making the original flavor disappear and homogenize.

Many, including the author of this humble book, believe that the best pizzas are found in New York and Boston. The best chocolate in the world is not made in Mexico, cacao nation, but in Switzerland and Belgium. Chocolate milkshake, typical of the United States, is now consumed in the entire world thanks to McDonald's.

Mediterranean diet, so perfectly described by Ancel Keys in the 1950's, is tempered by the lack of European postwar; current statistics show that Mediterranean people consume more meat, eggs, beer, and ice cream than what was previously believed.

The most expensive sushi and sashimi restaurant in the world—Masa—is not in Tokyo, Kyoto, or Osaka, but in New York City.

Most kitchen assistants in the best and most famous restaurants in this North American coast, it doesn't matter what type or ethnicity of food they serve, are Central American immigrants.

All cans of soft drinks, sodas, and beers in the world are round, so that they're easier to drink from; and all milk and juice cartons are square so they're easier to store.

The Cuban sandwich, prepared on flute bread with slices of ham, cheese, pork, pickles, and mustard, always measured about 25 centimeters. However, when the Cuban exile community in South Florida merged with the exaggerated American culinary style, the Miami-Cuban sandwich began to measure about one foot long, and in time and with these Cubans traveling back to their homeland, even the sandwich in Cuba began to grow excessively, completely abolishing the true Cuban sandwich.

The new concepts of deconstructive and molecular gastronomy extend throughout the world, applying pure sciences: physics, chemistry, mathematics, and the old and wise traditional food, leaving behind traditional ideas and grandma's dishes.

All of the above, a mere drop of water in the ocean of examples, illustrates a fact, at our unstoppable discretion, which is carrying food towards the same road of information, a group of "memes" without history within the increasingly smaller global village.

The stress of tourism, urbanism, media, social networks, migration movements, and the need to adapt and survive in hostile environments, is shaping a sort of planetary gastronomy that increasingly moves away from the old and popular culinary traditions of each country or region.

Identity is the set of factors (language, habits, rituals, etc.) through which people identify themselves, a set that is not static, but dynamic, especially when there is a marked social mobility.

This social mobility leads to multiculturalism, the famous melting pot is a typical example, and this multiculturalism is accentuated every minute through the so-called globalization.

Obviously, domestic foods carry a huge stress due to the relentless march of globalization.

On one hand, many countries seem to reinforce their national identities, including food, as a defense mechanism against absorbent multiculturalism, but if you look closely, you will see that this identity reinforcement involves a series of "identity assignments" that eventually end up distorting the original ethnicity.

Ethnic food is exported for competition, but its presentation format has yield, in order to sell the product, some ingredients are changed in order to please others, the product is also manufactured in other places to cut costs, and so on.

But the opposite is also true.

The media blitz—TV cooking shows, magazines, books kitchens, "new diets", etc.—and, above all, constant human interaction, are changing customs and traditions, tastes and needs of all people, regardless of the fact that some people, especially the elderly, insist on defending the traditional with great sacrifice.

Sociologist Giddens says that: "memory consists in the social organization of the past in regards to the present."

And globalization, as we understand it, has only accelerated this process of "past social organization".

Immigrants, which are many nowadays, bring their traditions with them, but the stress of the new environment, and the comparison, generally favorable to the new environment (there's a reason why they are immigrating, right?) start diluting or modifying these traditions. With time, these immigrants, and their descendants start modifying the customs and traditions they left behind, sometimes slowly and sometimes not.

A hundred years ago, there weren't many "contact areas" to different cultures. Take the example of a typical contact area to New York City, populated by Irish, Jews, Germans, Russians, Central Americans, Puerto Ricans, Europeans, Americans, and so on.

But today, these contact areas are almost endless. You can find a Central American or a Slovenian practically anywhere in the United States and an American in any city in the world. That's a fact of life.

The other problem, which is far from the concerns of this book, is the marketing and commoditization of tradition, which is a very interesting process, but not yet much researched.

Future (or present, why not?): A McDonald's, an Argentinean, a taco, a sushi, and a Korean restaurant, a Starbucks, a Lindt kiosk, a Cuban-Spanish post, a yogurt shop, and an international ice cream parlor, all on the same street of any city in China, Brazil, U.S. or Europe, are all clean, decontaminated, and certified.

Enjoy!

Cookers, cooks, and chefs

Cookers and chefs were and are our grandmothers, our mothers, our wives (and husbands, too), those aunts who lovingly prepare us coconut pudding or the kind neighbor who invites us for a barbecue and a beer.

Cookers feed the world and there are a billion of them.

All, absolutely all the cooks and chefs on this planet had, and probably have, a cooker at home, who obviously is practically never mentioned. Do they cook badly? Well... they actually cook for 99.98% of humanity. Now that we've paid tribute to them, let's now talk about the other 0.02%.

Cooks. Many of them would probably have preferred to be a firefighter or attorney; some did it, like Anthony Bourdain, who became a traveler, writer and television celebrity; or Vietnamese Ho Chi Minh, magnificent Parisian gourmet cook, as told, who ended up being a dictator and military conqueror of the French and North American.

Most cooks love what they do, but they lack resources or excessive ego to take the last step. They cook for money, just like a doctor or teacher charge for their work.

There are millions of cooks. Some do wonderful things in the kitchen, others, by necessity or apathy, end up behind a production line making fast food. And they grow and grow because we are increasingly eating out more. We are going up then.

Chefs, the real chefs, are just a handful. They cook—not much—they teach, open schools, and nowadays they handle the media anyway they want.

Let's recount some history.

There have always been good and bad cooks, but after the French Revolution, some of the best, the ones who cooked for the nobility, now exiled or beheaded, began to offer their services to hotels, to the new invention called "restaurant", or to the nouveau rich created by the Napoleonic empire. Those with more vision established classifications and titles to praise the way they manipulated and presented food.

This is how "Haut Cuisine" or "Great Cuisine" was born. It does not limit individual or ethnic creativity, but actually has more to do with using very expensive seasonings, elegant presentations, and complicated preparations. But most important, this type of cuisine has the urgent need of a leader who will direct the entire staff and make all the decisions as if he or she were a general in battle. This is a chef.

The first great chef, very talented in the kitchen and a genius entrepreneur, was Auguste Escoffier (1846-1935). He created memorable dishes and earned a lot, a lot, of money. He lead, like a field Marshall—he was a military cook in his youth—the kitchens of Le Faisan d'Or in Cannes (owned by him), the Grand Hotel of Monte Carlo, the National Hotel of Lucerne, the Savoy in London, the Ritz in Paris, together with Cesar Ritz, and the Carlton of London.

Just try to find someone with his resume! He brought to Western Europe the so-called Russian presentation—course by course with the approval of the diner—and the military discipline into kitchens.

Escoffier named his way of cooking, preparing, and presenting dishes "Classic Cuisine", a very intelligent maneuver that associated him, historically, to any respectable and dignified way of cooking, giving France a priority, a splendor, that nobody disputed in those days and which has endured, to this date, in united subconsciousness.

When carefully considered, Classic Cuisine was an important step–deconstructive, we would say today—in the dawn of ethnic food, since it labeled, as classic (and French), any intent of making gastronomy "elegant".

So much so, that around 1970, a rebellion of the chefs began, promoting a freer, more regional, less dogmatic, and recharged kitchen, named "Nouvelle Cuisine".

Among the great figures in this "revolutionary" approach to cooking, we find Fernand Point, Paul Bacuse, born in 1926, trained in a very harsh school from the black market and the French resistance during German occupation, Eugenie Brazier, a myth and the first chef to encourage her son to succeed as a chef in Disney World, Eckart Witzigmann, the first German to earn three Michelin stars, Alain Ducasse, a whole enterprise within him, generating more than 100 million dollars a year, Michel Bras, Gordon Ramsay, who has made art out of bad milk (and a good business), Heston Blumenthal, Juan Maria Arzak, Arguinano, Carmen Ruscadella, Ferran Adria, Herve This, Tatsuya Wakuda, Thomas Keller, according to him the best French chef in the United States, the Argentinean Francis Mallmann, and dozens more.

Television chefs have become celebrities today: Jamie Oliver, Nigella Lawson, Danny Boome, Tim Malzer, Rachel Ray, Bobby Flay and a long etcetera.

We can't forget the charming and afflicted mouse Ratatouille (a creation of scriptwriters Jan Pinkawa, Jim Capobianco and Brad Bird), touched by the gift of perfect seasoning, fighting against adversity and misunderstanding.

Even though you watch what you eat to avoid gaining weight, I share the opinion of celebrity chef Nigella Lawson (watching her is a whole party, it doesn't matter what she cooks!), who recently said in an interview: "I have no guilty pleasures; the only thing you should feel guilty about is not enjoying pleasure."

OBESITY.
AN ENEMY
WE MUST
DEFEAT

CHAPTER 22

The war against obesity

Losing weight, keeping in shape, being healthy: A haunting trio for millions of people in first world countries and a handful in others.

But... How?

The logical thing to do, in light of what we've read so far here, would be to: eat less, especially high-calorie foods, and move more, exercise, burn calories, but in real life, these two "simple" tasks may not be so easy to carry out.

Experience has taught us that the lack of time, work and social life, wrong foods, relaxing family life, the lack of willpower, and other obstacles make these goals very difficult to achieve, which should actually be carried out throughout our entire life in order for them to be effective.

This is where a multitude of shortcuts and tricks, which men have invented to achieve this longed objective, come into play, but, and this "but" is very painful, the adipocytes win the battle most of the time.

Let us briefly review some of those shortcuts and tricks, some in the spotlight and others already outdated (although trends always come back from time to time), but making it clear that both the prevention of fat and obesity, as well as the treatment of both conditions, should be studied, if the reader is interested, in textbooks written by trained professionals, or better yet, directly reaching out to these professionals.

Diets

On February 23rd, 1994, in a luxurious apartment in Manhattan, a 40-year old man was found dead. The autopsy revealed that the man weighed about 365 pounds and the cause of death was respiratory insufficiency due to morbid obesity, aggravated by the consumption of medications and drugs.

Thus ended the life of a brilliant nutritional guru, Dr. Stuart M. Berger.

After graduating with honors from Harvard and Tufts Universities, Berger soon discovered he had a talent for science dissemination and commercialization.

In rapid succession he wrote "Dr. Berger's Immune Power Diet" (1985), a book that narrates his own experience in losing weight, from 420 to 210 pounds in a few months; "How to be your own nutritionist" (1987), "The Southampton Diet", also in 1987 and three or four more books until just before his death.

At the time of his death, his medical practice, in a spectacular office with large windows facing Central

Park, was being investigated by the medical Board of New York for matters related to ethics.

In 2007, an interesting and unusual investigation was carried out about the basic words that appear in the cover of bestseller books from 1906 to 2006, precisely one hundred years. In first place was the word "Man", in the third place the word "House", and in fifth, the word "Sex/Sexual". What about in second place? Well, a word that appeared for the first time in 1922, and which never disappeared again: "Diet".

Back 400 BCE, Plato said that some doctors ordered diets, and therefore they were good for the slaves, and doctors who heard, reasoned, and explained; thus, they were good for the citizens.

You may find this hard to believe, but the concept of "diet", as we know it today, didn't emerge until the 20th Century. For thousands of years, doctors, barbers, bleeders, apothecaries, grandmothers, housewives, and anyone with good intentions could recommend a food or eliminate a food because it was too heavy or hard to digest, or even discover their beneficial action, such as the case of scurvy and lemons, but nobody we know had yet come up with prescribing strict and mortifying rules to acquire a new body image.

Men showed off their bellies with pride (as you can see with Napoleon and his hand over his abdomen) and women tried to compressed theirs with corsets and wires so that their hips and breasts could stick out, showing a fertility that sometimes translated into 10 and 15 children. This is how life was lived until the end of World War I.... when diets arrived.

Some progress and scores had emerged before. William Banting wrote a book in 1862, his "Letter on corpulence addressed to the public", which could have changed history, but his time hadn't arrived yet.

Monsanto introduced saccharine in 1879, but not for dietetic purposes but to offer a product cheaper than refined sugar.

Horacio Fletcher, known as Chew-Chew Man, around 1900, advocated the need to chew and chew foods until they were liquefied inside your mouth (John D. Rockefeller was one of his fans), but this trend didn't last long.

Greta Garbo was one of the first pioneers to take care of her body image through a healthy diet. "Nutritionist" William Hay, who enrolled Henry Ford as one of his clients, among others, created a system that did not allow the consumption of certain combinations of foods; for example sugars with meats, but he spoiled everything when he began to suggest a daily enema to eliminate toxics.

World War II brought the most appalling nutritional experiment ever seen in the history of humanity (it's embarrassing to say): concentration camps and the Holocaust. Millions and millions of men and women were exterminated by the absolute, or almost absolute, absence of food. But the most incredible thing is that some survived the experience. Stalin, with less propaganda, also eliminated entire populations in the same way, using subterfuge of collectivization.

These atrocities, as it occurs in the greatest catastrophes, increased the scientific comprehension of many metabolic and nutritional processes.

Since the 1950's, hundreds of diets and thousands of variations have emerged. The Shelton diet (1951) urges the victim, sorry, the "person", to starve... as simple as that. The diet of the Pritikin couple appeared in 1975— just in the peak of Jane Fonda's aerobic videos—and was based on eating very little and doing lots of exercise. The diet of South African Johanna Brandt was based on eating only grapes. The "Victoria Principal" diet, advocated by the Dallas actress playing Pamela Ewing in 1987. The Scarsdale diet, invented by Herman Tarnower in 1982 (shot dead a little later).

The tomato and onion soup diet was invented in a hospital. Rafaella Carra, Italian actress and singer, created the 8 a.m. diet. Eat whatever you want before that time. The hippies invented the grapefruit diet. Michael Montignac became rich in 1987 with the diet of not mixing proteins with carbohydrates. The potato diet is Spanish. The chrono diet, by Italian Mauro Tobisco (1991) alleges that carbohydrates are absorbed differently depending on the time they are consumed; this one is complicated but entertaining.

The so-called Mayo Clinic Diet, where you can eat only eggs, was a monumental fraud since it had nothing to do with the prestigious hospital. The Maple Syrup Diet became famous before disappearing into oblivion, after the film Dreamgirls. The blood-type diet was created in the 1950's by Peter D'Adamo, a neuropath.

The 1990's brought more elaborate diets, in a scientific point of view, which does not mean they are designed better or lack dangers. Some authors have named these diets or nutritional regimens as "heterodox" due to their low physiological observance.

Weight Watchers is a hypocaloric diet created by an obese housewife, Jean Nidetch, more than 40 years ago, with a strong group therapy (like alcoholic anonymous) becoming a gigantic commercial business.

The Ornish Diet is based on eliminating fat. The Atkins Diet has been the most controversial and the one which, generally, returns in disguise. The Zone Diet was created by Nobel Prize winner, turned industrialist, Barry Sears. We could be here forever mentioning each diet ever invented.

The reality is that all of these diets, without exception, trigger the so-called yo-yo effect. At the beginning, you lose water and muscle mass, then you start losing fat, but the adipocytes, which have a rather complicated hormonal and immunological defense system, begin to defend themselves and increase their reserve.

If a person consumes few calories and spends enough with exercise—for life, which is the key—fat is kept at a low, but this is not the case with diets, where you end up almost always on the rebound with greater weight gain.

This is the yo-yo of happiness and sorrows; lows and highs, but the highs are greater each time.

Aerobics and anaerobics. Physical exercise.

Striated muscle, which is what moves our skeleton and therefore our body, consumes a lot of calories. If the muscle works enough, it grows, hypertrophy, which in turn increases calorie expenditure, which together with a sufficient, but not excessive intake, should keep the person within normal weight limits.

Let's not get into the history of physical exercise here, which completely gets away from the subject of this book.

Let me clarify that there are no pure aerobic and anaerobic exercises. They both coexist in all shapes and exercise plans.

In 2005, Dr. Katherine M. Flegal published an interesting and controversial research which contradicts the popular notion that thin people live longer. There are other studies that prove that marathoners and long distance runners have lower life expectancy than the general population.

The subject, as already stated, must be studied in specialized literature and consulted with professionals.

Medications

Different medications have recently been used to treat fat and obesity. None of them, including those that increase people's "stamina" while trying to decrease their appetite, such as antihistamine, or those that work at the intestine level helping to limit the absorption of fats, have demonstrated any consistent effect. Actually, they have proven to have a large number of side effects that have forced some of them out of the market.

The truth is that we don't really have efficient medications, free of complications, to treat these conditions.

Nobody doubts that the day a medication is invented, equivalent to a vaccine or something similar to it, the history of obesity may undergo major changes, but until then, that moment is not envisioned.

Surgery

The history of surgical techniques and procedures that seek to improve body image deserve a book of their own.

We will only limit ourselves here to mention that they can be divided into three, quite defined, branches: 1—Plastic, cosmetic, reconstructive or aesthetic surgery, which generally eliminates portions of fatty tissues and excess skin, as is the case of the stomach and breasts. 2—Liposuction, invented in Europe around 1974 and improved a lot through ultrasound techniques, but also put in a microscope because of its high level of complications and negative side effects, and 3—The so-called bariatric, the only one really directed to attack a possible cause of obesity, which is the hyper ingestion of food, and projected to decrease the volumetric capacity of the patient's stomach, which can be achieved in different ways, none of them exempt of risks and complications.

They are all children of the 20th Century and their history is interwoven with the history of general surgery, anesthesia, and methods of preventing bleeding, infections, and other multiple complications, which, obviously, will not be discussed in this text.

Advanced research

Hundreds of clinical institutions and laboratories around the world work on various aspects of obesity.

There is evidence that the human intestinal flora may be involved in the increase, or not, of absorption of certain nutrients, such as carbohydrates.

The body's immune system may be intervening, along with adiposities, inflammatory reactions that make the body rebel against decreased caloric intake, thus perpetuating obesity.

The expensive tissue hypothesis, which refers to brain tissue, has been explored as a cause to the increasing need of extra calories in humans. Meanwhile, neuroscience is getting closer and closer to mapping the areas of the brain involved in addictions, among which is included, of course, food.

Understanding the operation of the genome, already codified some time ago, and their genes and non-genetic portions, of the metabolome, big picture of the body metabolism, of the proteome, a summary of all the proteins produced by the cells and their functions, and of the ambion, relationship between physiologist and environment, are gradually allowing an approach, ever closer, to understanding the latest causes of obesity.

The Pickwickian syndrome

Charles Dickens, one of the most important novelists of all times, was born in Portsmouth, England, on February 7, 1812.

Between 1835 and 1836, from his successive publications—just like soap operas in Latin America are produced—emerged "The Pickwick Papers", a novel where Dickens, who was just 20 some years old, demonstrated already his formidable literary technique and his enormous capacity of observation.

The main character in the book is Mr. Samel Pickwick, founder of the club which would later become famous with his name; however, Joe is the character who went down in history as the prototype of medicine and obesity. Joe was a glutton, extremely obese and constantly sleepy during the day because he didn't sleep well at night.

In 1956, 120 years after Dicken's novel was published, the investigation team of American professor C.S. Burwell published a scientific paper entitled "Extreme obesity associated with alveolar hypoventilation; a Pickwickian Syndrome".

They were referring to a 51 year-old male, weighing almost 300 pounds, who suffered insomnia, fatigue, sleep disorder and serious respiratory deficiency.

The medical term Obesity Hypoventilation Syndrome, due to morbid obesity and compression of the lungs due to thoracic and abdominal fat, came into view.

The Pickwickian Syndrome.

CHAPTER 23

Body Dimorphic Disorder

On February 4, 1983, a 32 year-old woman is declared dead at an Emergency Room of the Downey Community Hospital, in a Los Angeles suburb in California.

The immediate cause of death was acute cardiac arrest, but what drew the doctors' attention was the degree of emaciation (wasting) and edema she presented in her ankles and legs.

Liquid in the lungs, inflammation, and congestion of the liver and the spleen was found in the autopsy, as well as a distended digestive track with deposits of an already-dried substance, which is identified as tea leaves.

Talking with her family members, Dr. Edwards, the doctor in charge, finds out that the deceased took thyroid hormones in toxic quantities, despite the fact that she had no thyroid disease, ingesting emetine (ipecac) several times a day to produce vomiting. She also administered herself several enemas a day.

Of course she didn't eat either. From a scientific point of view, she died from a cardiac intoxication caused by the emetine.

But we all know she died of starvation.

The name of this attractive young woman, destroyed by her own hands, was Karen Carpenter (1950-1983), one of the most angelic and beautiful voices of U.S. pop music. Besides being a singer, she was also a fantastic drum player, and in time she confessed that she preferred the drums so she could hide behind them so no one would see her.

Together with her brother, Richard, she formed, in 1969, the duo "The Carpenters", with which they toured the world, topping, in just seven years, 16 songs in the national hit parade, selling, during the years they recorded together, 100 million records.

Karen, a tall beauty, never weighing more than 140 pounds, saw herself fat.

Perhaps she understood her situation intellectually, but when she saw herself in the mirror, no matter how thin she was, the image she saw starring back at her was that of an obese woman, deformed... and ferocious diets and extenuating exercise sessions renewed.

We were years away from neuroscience to begin revealing, by means of the research of "mirror" neurons and investigations with teams of functional magnetic resonance, the labyrinths of this disorder.

In 1975, at the peak of her career, she had to cancel a sold-out tour in Japan, where she had already obtained overwhelming success, due to her extreme weakness and her "concentration camp" image, as described by a journalist in a particularly vicious review.

She left two legacies when she died: her very beautiful voice and the publicity of her condition at death, until then pretty much unknown, recognized now as Anorexia Nervosa.

She would tell doctors who tried to administer intravenous nutrients into her system: "You win, I gained... pounds."

The relationship between obesity, anorexia nervosa and bulimia, two severe body image neuropsychological disorders, is not clear yet, but both conditions, perhaps two forms in one entity (nightly binging syndrome would be a third way) have greatly increased in the last fifty years, precisely when fat and obesity have skyrocketed around the world.

Either way, these are very serious disorders and often poorly understood by family members and those around the victims, who most of the times are adolescents or very young girls. The mortality rate of these disorders is very high and requires early and very specialized treatment to avoid complications, which often lead to serious organ deterioration and death.

And it is not strange to see the victims being related to the art and fashion industries.

Some cases of commercial modeling, of haut-couture houses, have reached to such an extreme that even the laws had to intervene.

Silhouettes of inconceivable thinness, "harmonic skeletons" as described already in the 18th Century, are the image of the supermodels who earn fabulous salaries at the cost of a strenuous and totally dysfunctional life.

Movie stars, television presenters, classic and contemporary ballerinas, are submitted to enormous pressures to keep lose and lose more weight.

Some can stand it, others can't, but the ones who survive pay a very high price to keep in shape.

The recent film "Black Swam", by director Darren Aronofsky, creatively deals with this problem, but the press, specially the tabloids, not so much.

As stated once by Walis Simpson: "You can never be too rich, or too thin."

¡You look like a Botero fat!

At the age of 24, things were not going that great for Fernando.

He was an artist, he knew that, but his contemporaries didn't know that yet. His uncle wanted him to become a bullfighter, but he responded to the claim by painting watercolors of bullfights.

He participated in some exhibits in his native city, Medellin, in neighboring towns and in the capital of his country, Colombia. He worked as an illustrator in a provincial newspaper, but his job caused him to be expelled from the art school he was studying for being "perverted".

He left to Madrid and enrolled in the San Fernando Art Academy. He continued to Paris and studied with unprecedented perseverance.

He diligently studied painting technique and art history, he tirelessly traced his teachers and visited museums.

He went to Florence. He went to Mexico.

And in the Mexican capital he had an epiphany.

He never explained in detail how he came up with his style: whether it was by playing with a special volume of the bodies, whether it was a form of sui generis surrealism, whether it was the influence of Spanish mannerism with Latin American sensuality, whether it was disproportionate expressionism.

Whatever it was.

He exaggerated a figure and represented in that figure any character of daily life: a priest, a teddy bear, a politician, a doctor, a car, a president (from any country), a construction worker, a housewife, a military, a little baby, a prostitute, a bicycle, an attorney, a crib, a bride and groom, a mailman.

"Boterism" was just being born, "The fascination with obesity", and with his new style, international recognition, success, and of course, the money that came with it all. Ah! And the enormous obese sculptures in different metropolitan cities around the planet!

He got to know glory and suffering. He won large quantities of money and gave away huge donations. It's a job that comes without rest. Everything is big.

If humanity disappeared and another civilization took over the Earth, Botero's sculptures would describe, exaggeratedly, the extinguished human beings.

Other great artists have painted, described, and illustrated heavily built characters: Trimalchio of Petronius, Ciacco of Dante, Shakespeare's fat Falstaff, Gargantua and Pantagruel Rabelais, Sancho Panza, Laurel and Hardy, Porky and Petunia, Marlon Brando, Orson Welles, Pavarotti, Elvis Presley, Tony Soprano (Gandolfini), but few have embodied obesity until it became its synonym.

You don't say Botero paints fat people anymore... no...

Now you say: "Botero's fat people"!

CHAPTER 24

Is obesity a disease?

The Centers for Disease Control and Prevention in Atlanta (CDC) reports that some 73 million Americans are overweight or obese. The University of Puerto Rico in Rio Piedras reports, in a very interesting study, that approximately 60% of the island's residents suffer from the same condition.

The World Health Organization (WHO) states that people who are overweight or obese more than triple the number of inhabitants of the planet who are hungry and malnourished. All these institutions believe that obesity is destructive, decreasing quality of life and threatening the future of humanity, among other reasons, because of their genes and their negative effect on the normal physiology of an individual.

But not everyone agrees.

Many researchers and communicators believe that this obesity epidemic is due to the huge increase in the consumption of high-calorie beverages and the increasing decrease of physical activity; therefore, it is a matter of

personal responsibility and not a disease acquired by
uncontrolled external agents.

None of these researchers and disseminators argues that
obesity can trigger real, and very serious, diseases, such
as diabetes mellitus, hypertension, pulmonary disease,
joint diseases and injuries, and even certain cancers, but
without being itself a pathology. Where is the truth?

So let's review the arguments. Those who argue that
it is NOT a disease claim that if overweight or obese
individuals decided to stop consuming large amounts of
calories, empty or not, and began to move enough to
spend their excess calories, they would stop being obese.

They point out that children become obese due to poor
parental education and poor school supervision; in
addition, of course, we have the media blitz propaganda,
which is ultimately an economic and social problem. As
senior researcher J. Justin Wilson says: If obesity is a
disease, then it is a disease of society, and not just the
person.

Those who defend that it IS a disease, point out that:
Obesity has an undeniable genetic predisposition, which
explains why there are sedentary people who eat fast
food all day who never get fat.

Obesity is also a compulsive psychological disorder that
can be treated medically.

As Professor Scott Kahan (Johns Hopkins University) states:
obesity is a deregulation of the control of caloric storage
at the cellular level, in the same way that hypertension is
a disease of the arteries.

Obesity leads to cellular changes, inflammation, increased circulating fats in the blood, especially the "bad" ones, hormonal imbalances, atherosclerosis, certain cancer risk, serious problems with self-esteem and increased depression, hypertension, pulmonary air flow hypertension, sleep disorders, injuries to bones and joints, difficulty moving, etc.

In short. Certainly, there is a socioeconomic factor in obesity, but associated to a genetic disposition that makes its emergence possible.

For the author of this book, a Doctor by profession, obesity IS a disease, but increasingly facilitated by the distortion and dysfunction of today's lifestyle.

However, the topic is not closed to discussion.

What do you think?

Good umami, bad umami

Taste is more than flavor. In school, we are taught four basic tastes: sweet, sour, salty, and bitter, which the progress of microanatomy and physiology associate to several different types of taste buds scattered on top and on the sides of the tongue.

Taste is much more than the simple mix of flavors, since it depends, 70% or more, on the nose and the brain. That's why when we have rhinitis, all foods taste bland, or when a person has brain injury, all tastes are lost.

For scientists, pungent, astringent, and creaminess are not tastes but a chemical effect they call "chemesthesis". A good example of this is menthol, which leaves for some time, a cold sensation throughout the mouth and nose, and is not a taste.

However, a Japanese scientist came to complicate everything. In 1908, more than 100 years ago, this professor from the Imperial University of Tokyo, Kikunae Ikeda, who loved the taste of raw fish and seaweed he ate almost every day, came to the conclusion that these foods were not sweet, nor acid or bitter, and even though they had some "saltiness", they weren't salty either. So what did they taste like?

He analyzed in his lab hundreds of substances extracted from foods that he liked very much, until he came upon the salt of an amino acid called glutamic acid, which forms part of the protein of these products and of many others, such as meat, some cheeses, tomatoes, and different vegetables. Ikeda, who was a good chemist, called this substance monosodium glutamate.

He even went a step further and synthesized a certain amount of the substance and he added it to things that had no flavor, such as unsalted hot water with noodles, and he excitedly discovered they became "tasty"! And since "tasty" in Japanese translates to "umami", he named this fifth taste sense, activated by a monosodium glutamate, umami.

And gradually, everyone accepted the fact that besides the traditional tastes, umami was a fully rightful taste that answered many culinary mysteries.

About 93 years later (2001), researchers from the University of California identified some taste buds located in the tongue's central area that specifically perceive the umami taste.

They also discovered that other proteins, besides monosodium glutamate, activate these buds.

It was also discovered that the number of buds varies from person to person. There are people who have many of them and they are called "supertasters" and others have very few of them, which doesn't allow them to enjoy the delicatessen of the gods.

If you decide to make fun of one of these unfortunates who seem content when eating bland foods, think that because of convoluted genetics, this person may lack sufficient umami in their taste buds. Be kind.

How about obesity? In the last decade, several studies appear to have demonstrated that monosodium glutamate and other similar proteins have a multiplying effect, some say addictive, regarding appetite, which can lead the brain to momentarily lose the mechanisms that limit the amount of food ingested.

The phenomenon is explained very well with the example of french fries: the more we eat, the more we want to eat, especially served with ketchup!

Most sauces, not to say all, are basically umami. The most umami of all is the soy sauce we immeasurably pour over our Chinese food—or at least what they've made us believe, in this side of the world, is actually Chinese.

This sauce is closely followed by ketchup, which is the most consumed sauce in hundreds of countries around the world, and most consumed by children and adolescents, especially when generously mixed with any type of food, especially junk food.

So let us review the whole picture: The food industry knows that the umami taste has addictive factors in children and adolescents. Almost all fast food is umami, whether it is good or bad, children and young people prefer the known, which is umami, to the new, neophobia, which may be of greater nutritional value, but is not umami. The umami taste of these foods is fixed in the tongue and the brain (aftertaste) and becomes a habit: addiction.

Here we close the circle that can help explain, among other factors, the obesity pandemic.

CHAPTER 25

The future of obesity

Let's begin this last chapter with an affirmation, which is at the same time a definition.

Obesity, or in layman's term, fat, is the physical expression of a predisposition that is implicit in the human genome, or of many humans, but also in the genome of other animals, because our pets, our little dogs and cats, can also get obese when they lavishly live with us, just like pigs, cows, chickens in a farm, and lions and monkeys in a zoo, and is manifested when the environmental conditions facilitate its emergence: sedentary lifestyle, excess of high-calorie foods, addictions, stress, etc.

Obesity is already a pandemic and the cost of other related diseases rise and multiplies to unthinkable values.

The U.S. Army reported that 25% of male and 40% of female in drafting age are discharged due to obesity.

Stretchers for obese people are essential in first world countries. Airlines spend as much fuel for the average increase in weight of their passengers as for the luggage, and tourist class seats are increasingly getting narrower and shrinking in order to accommodate the expected number of passengers.

But the other side of the obesity pandemic and the spiral of costs is the extensive and substantial research done on the subject.

Everyone knows that there is huge financial gain and glory in the elimination, or at least control, of obesity and fat. Countless researchers, as we have noted before, struggle to find the organic causes of this condition.

If medications, such as Viagra, antidepressants, and anticholesteremics have given billion-dollar profits to pharmaceuticals, imagine how much more money products that were really effective against these conditions would bring in.

Perhaps this future is not that far away.

On the other side, sectors of the population have started to confront the undeniable social and employment discrimination that obese or heavy people suffer. "Fat Pride", "Size Acceptance", "Fat Liberation" movements, among others, are correct to demand good treatment without discrimination towards the obese. These movements will eventually grow in the future.

The future, good or bad, will decide on its own, but any indolent and non-proactive attitude related to this huge problem of obesity pandemic, whether at personal, family, scientific, social, school, or governmental levels, is simple irresponsible.

It would be fantastic to have books, such as this one, that do not need reediting, except to tell everyone that, finally, the obesity pandemic it is over.

What do astronauts eat?

The first few trips to space, in the 70's, had no issues from the nutritional point of view, because they were short, just a few hours or a day.

However, as the flights, from Russian cosmonauts as well as U.S. astronauts, became longer, the issue of "what to eat in space" became an extremely significant concern to the planners on Earth.

The first two challenges to solve were gravity, meaning, the lack of gravity inside the spaceship, which made everything float, and the second was the packaging and preservation of food.

At the beginning, they solved both problems by dehydrating foods, in other words, they eliminated all of the water contained in them, turning them into powder, packaging this powder then into little plastic bags, or turning them into paste and putting them inside tubes, much like toothpaste.

Astronauts didn't like the taste of these foods, some of them returning to Earth after their missions having lost a lot of weight because they preferred not to eat, disobeying the orders of superiors and commanders.

As experience was gained, the quality of the foods improved a lot, and they became more balanced, digestible, and nutritional.

In the 80's and 90's, technology in the preparation of foods as well as in the miniaturization of equipment, allowed the astronauts to eat hot or cold meals, according to their needs and tastes. They were able to use sauces and other seasonings that made their food taste and look better.

Long trips and living in International Space Stations from different countries have produced enormous amounts of information about this subject. Various research labs, not only in the United States and Russia, but also in countries such as China, Japan, France, Israel, Canada, and many others, have enriched the knowledge of nutrition as well as of preservation and presentation of foods under the conditions of weightlessness and isolation.

The new challenge resides now in trips, which are already being planned, to distant places, such as Mars and missions of another type, such as the capture of a small asteroid.

Nutritional requirements of people in space change and modify even more as weightlessness goes on for months, or even years.

For example: Human caloric needs decrease due to muscle decrease (which constitutes, per se, a huge physiological problem), but the need of calcium increases due to the loss of this mineral in the bones, which creates, in turn, the possibility of calcium deposits in the kidneys, producing kidney stones.

We have just mentioned a few of the problems confronted by doctors and physiologists who dedicate themselves to this interesting and very new branch of science.

The fact is that these and many other problems must be resolved and surely would be resolved, bringing in new ideas and knowledge that can affect, for the better, other sectors of human knowledge.

APPENDIX

The American Medical Association (AMA), in June 2013, recognized obesity as a disease and not just as a simple risk factor.

This statement, while obvious, is very positive as it will deal much more effectively with the millions and millions of Americans who suffer from this condition, now admittedly morbid.

It will also facilitate handling overweight youth by avoiding the fear of creating misunderstandings and disparages when referring to a specific physical condition that was not previously taken as a pathological entity.

It will also facilitate the increase of pressure on pharmaceutical companies to find new drugs to fight this pandemic and on the manufacturers and packagers of foods to control their relentless hypercaloric offer.

That's good news!